The Modern Girl's Guide to Sticky Situations

Also by Jane Buckingham

The Modern Girl's Guide to Life

The Modern Girl's Guide to Motherhood

What's Next

The Modern Girl's Guide to Sticky Situations

Jane Buckingham

AVON

An Imprint of HarperCollins*Publishers*

The advice contained herein is for informational purposes only. Please consult your medical professional before embarking on any medical program or treatment and see a financial advisor before making any investment decisions. The author and the publisher disclaim any liability for any losses or damages resulting from the use of the information contained in this book.

HarperCollins books may be purchased for educational, business, or sales promotional use. For information please write: Special Markets Department, HarperCollins Publishers, 10 East 53rd Street, New York, NY 10022.

FIRST EDITION

Based on the Design by Judith Stagnitto Abbate/Abbate Design

Library of Congress Cataloging-in-Publication Data
Buckingham, Jane, 1968–
The modern girl's guide to sticky situations / Jane Buckingham. —1st ed.
p. cm.
Includes index.
ISBN 978-0-06-177635-9
1. Etiquette for women. 2. Women—Life skills guides. 3. Women—Conduct of life. I. Title.
BJ1856.B79 2010
395.1'44—dc22 2009040343

10 11 12 13 14 OV/RRD 10 9 8 7 6 5 4 3 2 1

To Marcus, Jack, and Lilia, who make even the stickiest of situations sweeter

Contents

Introduction

I am no stranger to sticky situations.

I could fill the pages of *Vogue*'s massive fall fashion issue with all my cringeworthy foibles. There was the time when my good friend in high school slept with my boyfriend—because I wouldn't (and then I had to determine whether to uninvite her from my sweet sixteen). Or the time I angered a group of very political—and hormonal—women with my advice on breast-feeding (I believe that whether a woman chooses to breast-feed is up to her and no one else).

One of the stickiest situations in which I've found myself concerned—of all people—the country's florists. Gentle, creative types who spend their days in gardening gloves, listening to classical music, seem unlikely to cause trouble, but aren't the toughest situations often the most insidious ones? After all, trouble is only trouble when it catches us unprepared.

Here's what happened: I wrote a book called

The Modern Girl's Guide to Life, which offered lifestyle tips aimed at turning the domestically challenged into domestic (demi)goddesses. In the book, I suggested bringing something other than cut flowers as a hostess gift when attending a party, since cut flowers require the hostess to stop whatever she's doing and come up with a vase and some water to put them in. Who needs yet another task in the middle of all the other pre-party panic, when the hors d'oeuvres are burning and the ice cubes are melting and the guests are ignoring one another?

Shortly after the book was published, I received a letter from the florists' association. "Dear Ms. Buckingham," it read, "On behalf of the 45,000 florists of America, we're very upset with your suggestion." Actually, "upset" was putting it mildly—they were furious. They went on to say, "As florists, we are simply trying to bring a little bit of joy into people's lives!"

You might think I would dismiss such a note with a laugh, but I really felt terrible. I had never meant to offend florists! I love flowers! And despite the fact that my book was selling well, I was still surprised that anyone had actually read it! Could what I had to say actually hurt people? Once I'd choked the sickening feeling in my stomach with a handful of Tums, I forwarded the letter to my husband, Marcus, along with a note: "Do you think I should send them some flowers?" Instead, I sent a sincere, handwritten apology.

Humor is often the best way to deal with an

uncomfortable predicament, and of course time heals many wounds. But sometimes a girl needs to be told exactly what to do in a difficult situation. And that's where this book comes in.

You know that moment when you just want to scream, "Help"?

When you wish you could light a flare and wait for an airplane full of Coast Guard studs to shimmy down from a helicopter and rescue you? We all know sometimes a girl just has to face things on her own. But what if she just . . . can't?

This is the book to grab when you're pulling your hair out, screaming like a banshee, blubbering with frustration, or hyperventilating into a panic because you *just don't know what to do*.

Fear not; no problem is too big—or too small—for the Modern Girl. Catch your mate's man cheating? We've got the answer. Soufflé imploded? You can still dish it out with style. Skeevy authority figure hitting on you? We'll help you ward him off sans drama.

This is the book to turn to when you have nowhere to turn, or are simply too embarrassed to turn anywhere (publicly). A book that answers the sorts of question you'd only feel comfortable posing to your best friend, but which only your mom, or grandma, or gynecologist, or cleaning lady—okay, any woman who secretly intimidates the hell out of you—knows the answer to.

We live in a culture of prevention. Take your vitamins! Do your sit-ups! But we're also a country of procrastinators. What a lethal combination.

This contradiction between nature and culture leads to lots of everyday emergencies we don't feel qualified to handle, because we're supposed to have prevented them in the first place! So we panic.

Of course everyone means to do a little spot cleaning each morning after breakfast, but what if you just haven't felt like it for a week or so and then—surprise!—your boyfriend's mom decides to make a special appearance? (Someone needs to explain to me why it's so much easier to commit to wearing clean underwear than to spraying Tilex daily.)

The good news is this: Most hideous predicaments are not the end of the world. This book is a collection of quick fixes—an industrial-sized crate of metaphorical fire extinguishers, if you will. It's not meant to replace responsible behavior (for some idea of what that entails, check out *The Modern Girl's Guide to Life*), but to be a stack of get-out-of-jail-free cards for the deserving gal with the best intentions and a moment of bad luck.

Sticky situations come up whether you're toiling away in your cubicle or tooling around the house, trying to be the life of the party or trying to throw the party of a lifetime. And so this collection of solutions is a mix of silly and serious, just like life.

The Modern Girl's Guide to Sticky Situations is organized by situation, so solutions are easy to find. It also has a pretty amazing index in the back, so you can find your particular dilemma no matter where or when it happens. And just so you know, there is no judgment here. We know one girl's drama is another one's daily life; trauma is subjective. Isn't it just the worst when you finally muster up the courage to ask a friend for help and her steely response is, "What's the big deal?" (Cue eye roll.) Now you won't have to worry about being judged for hating the five pounds only you can see. Or being obsessed with the invisible mold in your bathroom, the stretch marks you swear are all over your boobs, or the way your boss looks at you when he's talking to his wife.

You might be wondering how one woman can be an expert on both babies and bay windows. Unless she's Rosie the Robot from *The Jetsons,* she probably can't. But I've never once called myself an expert. Rather, I fancy myself a filter—an information curator of sorts. My office is filled with sticky notes covered in info tidbits—the same tidbits that fill the margins of this book. The ideas in this book come from experts, friends, experience, the process of writing my first three books and running my trend-forecasting company, and my own bad—and good!—luck. It's as much for myself as for my readers that I decided to gather all the lifesaving information I've come across over the years into one place.

So think of this book as your own personal stash of missile defense codes, your vial of snake venom antidote, your get-out-of-jail-free card. Just when you think all is lost, here comes the *Modern Girl's Guide* to save the day.

PART 1

Sticky Situations in Love

This is one area in which I have far more expertise than I would like. There was the time my "true love" was only using me as a distraction while his girlfriend was in Europe for the summer, the time my best friend's boyfriend declared his secret love for me, the time I found out the guy who'd said he was going out to put money in the meter had really ditched me—and stuck me with the check (I used to tell myself he'd been hit by a Mack truck, but I realize now I was delusional). Best of all, one guy told me he was gay so he wouldn't have to date me any more—but he subsequently went on to date several of my best female friends.

The good news is, after kissing my share of not only frogs but also slugs, weasels, vermin, and downright asses, I did find my prince—in the most unlikely of places! (A blind date in Orlando.) So whatever your status—searching, settled, or satisfied—don't worry, there's a sticky here for you.

Dating

You're having a fantastic time when all of a sudden your date says something so offensive you're pretty sure you misheard him. But then again, you're pretty sure you didn't.

First, make sure you're understanding him correctly. If you don't know how to phrase this question, say, "I just want to make sure I'm understanding you correctly. Did you really say all orange foods should be banned?" (Insert whatever's relevant.) If you're lucky, he'll deny or qualify. If he persists in the opinion that turns your stomach, call him on it—but be demure. Tell him you disagree, but try to keep emotion out of it—at first. The more well-thought-out and less knee-jerk an argument, the more sense it usually makes to a man. (Sorry to generalize; it's biological.) However, if he does hold beliefs that are offensive to you, calmly make it clear that you disagree with him, then start wrapping up the date. It's not your job to change his narrow little mind.

You inadvertently insult your date— or perhaps worse, he has the gall to insult you!

Nervousness has been the cause of a thousand gaffes. As soon as you realize your mistake, apologize, blame your wacky sense of humor, and ask to start over. A little bit of backtracking will usually earn you forgiveness.

If your date insults you but it's clear he didn't mean to, either let the moment pass without acknowledging it or make a joke, like "But you like *my* pink hair, right?" If he really did insult you big-time, end the date quickly. You deserve better!

Your date shows up drunk.

Is he tipsy or toe up? Nervous daters have been known to overdo the liquid courage. You can always hit the bar and catch up with him (suggest he match your cocktails with cups of coffee), or join him in a carb-heavy meal—by the time you get to your entrees, he'll probably be sobered up (and a bit embarrassed). Now, if your squire is completely smashed, put him in a cab and wish him well. You are not the local chapter of AA.

You were married in your early twenties—for seven months. Is this a big revelation or is it just another past relationship?

It's best to share this story when you get to the dating history stage of the romance. Go down the litany of exes, and add, "And, well, I was actually married for a little while right after college." Most likely it will be seen as just what it was: a youthful decision that went awry.

You've got a big secret (kids, a messy divorce). When do you share it with your new interest?

The sooner the better. You don't have to wear a name tag that says, "Hello, my name is Recently Separated," but you should share your status with anyone you are considering a relationship with. It doesn't have to be a dramatic revelation—casually show him pictures of your kids on your iPhone and explain the custody arrangements. If he can't handle a complicated situation, it's best to find out earlier rather than later.

You're not a one-night-stand kind of girl, really, but somehow last night you just wound up, well, you know, having a little too much fun. So how do you deal with an awfully uncomfortable aftermath?

The way I see it, there are two possibilities here: Either you're wishing he'll call or you're praying he won't. If you like him, the best thing to do is wait—for the first forty-eight hours. If he hasn't called by then, it's not a great sign, but there's still hope that logistics got in the way. Call or text him that you enjoyed yourself (wink wink) and would love to see him again. Short, sweet, declarative. If you get no response, walk away. You don't want to have to convince anyone to like you this early in the game; even if you end up together, you'll spend the whole relationship wondering whether his feelings are real.

If you hope never to see him again but he gets in touch, be kind. Think of how you'd want to be treated. Don't disappear; pony up and say the way you acted was out of character—maybe even that you're a bit embarrassed. Just be honest. Come to think of it, this isn't the worst way to act with someone you do like, either.

A friend's crush goes after you, even though you have no interest and haven't done anything to provoke it.

Turn him down the way you would any other pursuer you aren't into—politely and firmly. And if your friend has caught wind of the situation (or, God forbid, was present while he made his play), make it clear to her that you would never angle for her dude. Then step aside and let her continue working her magic.

A friend's crush goes after you, and you think he's pretty cute.

Make no moves without checking in with your lady friend. If this gentleman has been a long-term project for her, you may just want to bow out preemptively no matter how fine he is. But if he's a recent romantic goal, she'll probably be okay with handing him off. Either way, what she says goes. Hos before bros, yo.

You want to Google your date.

Some of you may be thinking, *What's sticky about this? Doesn't everyone Google their dates?* Sure, but how many of us are thrilled with what we find? My advice is simple here—only Google him if you're prepared to find out some pretty weird stuff. Maybe he wrote an editorial about why he doesn't believe in monogamy—or why he does believe in Santa Claus. Maybe he's been in the local paper for all the wrong reasons. Maybe the *New York Times* ran his first, second, and third wedding announcements. My feeling is that total access to information sabotages the natural process of courtship. He should be able to tell you about himself in his own time. Finding out too much prematurely can lead you to draw the wrong conclusions—after all, you never know who shares his name.

You're on a blind date that seemed promising at first, but now there's zero chemistry and you've barely gotten through the appetizer. Do you fake an emergency, or do you stick it out, knowing TiVo has you covered?

Is the lack of interest mutual? Take his temperature—when you flirt, does he flirt back? If he's into you but you're not into him, is it better to be honest or to save his pride by making up a white lie such as "I'm about to get back together with my ex-boyfriend"? Being honest—acting available but disinterested—may backfire and lead him to pursue you, because we all know men love a challenge. But you must never make yourself unappealing or too unavailable, because you never know who could lead to someone else. What a dilemma! Just wrap the date up quickly. If he tries to order dessert, demurely protest by saying you have an early morning meeting and pray he gets the signal.

When I was on the market, I used to always schedule blind dates as drinks, never dinner. That way, if your chemistry is as cold as the North Pole, you can call it a night. If it's as steamy as Rio, head out to dinner.

If he calls you for another date, don't be tempted by the lure of a free meal, good concert tickets, or just one fewer night at home watching *Mad Men*. Just say, "I had a really good time, but I think maybe we should wait a little while before we go out again." When he tries you again, tell him you're really busy but that you have a pal you know he'd like, and hook them up. Maybe he'll return the favor.

Dear Jane,
I have a simple question: Who pays? When is it okay to offer—or not to offer—to split the bill?

Always have the means to pay for yourself, even if you are out on the town with an investment banker with cash to burn. Offer to pony up for your half; more often than not he'll refuse. If you can't stand not going dutch, you can always offer to pay for after-dinner drinks. Switching off who picks up the check is a more elegant solution than splitting the bill down the middle and counting out cash.

You agree to a second date even though you know you're not into him.

As tempting as it may be to keep your options open, don't give false hope to someone you have no interest in. "A girl's gotta eat" is no reason to reprise a mediocre date. Give him a call and say, "Look, I've been thinking about it—I don't think we're a good match and I don't want to waste your time." If you genuinely want to move things into the friend zone, be up front about it. When setting up the date you agreed upon, say, "I really enjoyed spending time with you, but I think we would work best as friends." Back this up with a suggestion of a fun—and platonic—activity like bowling, and be prepared for him to pass up your proposal.

You have some shameful physical secret and you're terrified he's going to find out.

What, do you have a vestigial tail or something? Women always have a laundry list of things they hate about their bodies; if you are going to have a real relationship with someone, you are going to have to risk exposing the parts of yourself that you don't like. Whether it's a scar, a weird birthmark, or an extra toe, it's likely that he's going to love it as much as he loves the rest of you. That is, if he even notices it. And look, if you do have something major, it's part of you, part of what has created who you are—made you stronger, wiser, and deeper. Don't be ashamed; be proud. If he doesn't love it, and you, he doesn't deserve you.

You want to take a friendship to a romantic place, but you're not sure your pal feels the same.

Ah, here's a sticky that launched a dozen teen movies. Well, you can either throw a John Hughes film festival and then try various misbegotten hijinks inspired by Molly Ringwald, or tell him honestly about your feelings and accept his response. If he's not interested, you should know that there are lots of reasons a man might not want to take the relationship to the next level. It really doesn't matter what they are or if they make any sense to you. The

Things That Make a Terrible First Impression

- **Neglected pubes.** Nobody's saying you need to go completely bare, but you must pay some sort of attention.
- **Stray hairs on your nipples.** If we're talking something from *The Planet of the Apes,* you may want to talk to your doc, just to make sure your hormone levels are okay, but if it's a few stray hairs, that's what tweezers are for.
- **Body odor.** Whether it's coming from your armpits or your Frye boots, keep it clean, ladies.
- **Potty mouth.** He may eventually think the fact that you speak like a sailor is *très* adorable, but let him become acquainted with the rest of your vocabulary first.
- **Old or stained lingerie.** What if you get into an accident? Or a tawdry affair?
- **Bad breath.** Scrape that tongue, brush frequently, and for goodness' sake, don't whisper.

- **Chin hairs.** The best place to reveal them in all their glory? The car mirror. I keep a pair of tweezers in the glove box.
- **Dirty fingernails.** Who are you, Dennis the Menace?
- **Dry, cracked lips.** Do you really think he wants to make out with crispy sheets of peeling flesh?
- **Spiky legs.** Either shave or let it grow, but that in-between stage feels on his legs like his beard feels on your face.
- **Sandpapery heels.** If they'd put a run in a pair of pantyhose, it's time for a pedicure.
- **Too much makeup—especially in bed.** He'll feel as though you're hiding something.
- **Wigs when you're not sick.** Rock the hair you have, even if it's a wispy pixie cut. Worked for Mia Farrow.
- **Your ex-boyfriend's clothing.** Sporting someone else's team jersey or pajama top isn't going to make him give you his.

bottom line is that when he imagines being with you more intimately (and trust me, he does think about these things), he pauses and then says to himself, "Nah." Consider taking a break from the friendship until your emotions calm down.

On the other hand, if he confesses his undying love for you and you feel differently, let him down easy. Be clear about wanting to keep things completely platonic, and follow through. Don't play with his feelings to pad your ego—it's not fair to use him as a substitute boyfriend or devoted eunuch. Again, you may need to cut down on the time you spend together, at least until he finds someone new to crush on.

Now, the stickiest of these situations is when friends just have sex, or when friends think they like each other, sleep with each other, and one changes his or her mind. When this happens, chances are, the best thing to do is take a break from each other. Recognize that good friendships are almost as hard to find as good romances. If your love match wasn't meant to be, your friend isn't punishing you, and it's not your fault—so why throw the baby out with the bathwater? Be friends, but *never* sleep together again unless the friendship has turned into a serious relationship. And while there might be a rare exception, I've found that "friends with benefits" tends to be a lot like health care—the coverage isn't there when you need it most.

You're dating someone you're crazy about, but nothing's official—should you change your relationship status on MySpace, Facebook, and Ning?

This is a delicate process with several different stages. First off, when you start dating a guy exclusively, you might consider removing the single status from your profile out of respect for what you've got going. Warning: doing this will cause a pileup of curious pals ("Woo-woo!" "Who is he?" etc.), so be sure to delete the move from your newsfeed. Once you and your man move into boyfriend-girlfriend territory, it's time to shift (preferably in sync) to "In a relationship." Finally, you may choose to link your profiles, which involves one of your sending an invitation to the other, much like a friend request. This is best reserved for solid relationships—if you're in the breakup/makeup cycle, it's smarter to reveal as little of this as possible to the coworkers and cousins keeping their eyes on your bidness.

You don't know how to introduce him when he's not your boyfriend but more than a friend.

Skip the descriptor entirely and say, "I'd like you to meet John." The fact that you don't refer to him as a friend or boyfriend is typically telling enough. Your body language will likely fill in the blanks for

- **Excessive criticism of other people—or yourself.** A huge turnoff—he'll likely just be imagining what you say about him behind his back.

What He Really Means When He Says . . .

He says: "Let's just be friends."
He means: I never want to see you again.

He says: "I value our friendship too much to date you."
He means: I'm not attracted to you.

He says: "I love you but I can't be with you."
He means: I'm seeing someone else (or) I'm not ready to commit to a relationship.

He says: "You'll always have a piece of my heart."
He means: We did some really embarrassing things in bed and you'd better not tell anyone.

He says: "It was just the wrong time, wrong place."
He means: I was drunk.

people. If that's too ambiguous for you and you want to mark him as your man, call him your date.

He says: "I'm sorry. Can we try again?"
He means: My girlfriend broke up with me.

He says: "I'm just not ready for a relationship right now."
He means: You're not the one.

You can't get that guy out of your head—and there's no room for anything else.

Ah, sweet infatuation! First off, enjoy it, 'cause it's not going to last long. Every new couple should have some time of mutual obsession—there's nothing wrong with getting caught up in new love as long as it doesn't keep you from eating and going to work.

If the man on your mind is an ex-boyfriend or a crush that never came through, give yourself a couple of weeks of mooning around and then pledge to pull through. Work is a great distraction—ask for extra projects, focus on a complex task, and put in overtime. Now is not the time to take up mindless activities that will give you time to daydream. Before you know it, you'll barely remember the guy—and you might get a promotion.

You want to know if it's okay to start calling him your boyfriend, but you don't want to seem needy or clingy.

If you've been dating for a while, he's probably got the same question on his mind. Ask him for what you want in a lighthearted way: "So, are we going steady or what?"

You have a one-night stand with a random guy while getting to know someone you might really like.

Technically this is fair game. Until you are in a committed relationship, you don't have a responsibility to be monogamous. That said, a lasting relationship can be worth sacrificing some easy sex for. You might choose to skip the hookups until you see where things are going with Bachelor #2.

You're dating a guy who lives with his parents.

When the economy is crappy, it becomes more and more common for someone to be bunking with his folks. It doesn't have to be a deal breaker—you just need to distinguish between Peter Pan syndrome and a genuine time of personal transition. Is he focused on working (or finding work, if he's been laid off) or school? Is he taking care of ailing parents? If so, you might want to weather the discomfort of shagging on the top bunk. However, if he's just sitting around the old homestead eating Mom's meatloaf and playing Second Life, peace out and leave him to his arrested development.

You're dating more than one guy and don't know whether to be honest about it.

Until you've discussed a commitment, you're technically free to date around sans guilt. But that doesn't mean your gentleman friend won't have hurt feelings when he finds out he's not the only name on your dance card. If you want to segue into a relationship with one of the dudes, then it's easy enough to phase out the other one(s) without full disclosure. However, if you plan on man juggling for a spell, you should make it clear that things aren't exclusive. It's best to bring this up when setting up plans, especially if he starts making assumptions about your time. Say, "I really enjoy spending time with you, but I feel like I should let you know that I'm seeing other people." Then it's up to him to decide if he's okay with the casual thing.

Everything is going great. Really. So you're pretty sure this is where you'll screw it up somehow, like you always do.

Breathe, breathe, breathe—it's natural to feel fear or anxiety in the early, ambiguous phase of a relationship. Use your girls as your sounding board and take the time to examine your worries. Are you afraid of being close to someone? Of losing your freedom? Of opening yourself up and (potentially) being rejected? Feel free to work it out with the ladies, but for the meantime, hold back from sharing your fears with your beau. Sometimes we want the object of our affection to see us cry and be vulnerable because we want to be accepted or protected. But hold out a little longer. In the meantime, friends, family, and therapists are there to sop up the crazy.

Having said that, be careful of the faux friend. This is the girl who likes to pretend she's got your best interests at heart—but really doesn't. She's usually single and a bit bossy, and she doesn't really want anyone to be happier, prettier, skinnier, or tanner than she is. She may not even know she's doing it, but she'll steer you in the wrong direction and give you bad advice. Also, this is not a public poll. Talk to one, two, or at most three of your closest friends—don't let word get out that you're worried, because you don't want word to get back to him.

You made the big proposal! Okay, not *that* proposal—the one about moving in! He doesn't want to. Ugh. Should you break up just to save face? And by the way, why *doesn't* he want to live with you?

What are your reasons for wanting to move in together? Do you want to bump your relationship up a stage, or do you just want to save cash and com-

Things Guys Don't Mind Nearly as Much as You Think They Do

- **Cellulite.** Your cellulite is a million times more visible to you than it is to anyone looking at you, because of the way the light reflects from your eye to your thigh—and the fact that you're likely obsessing over it in a dressing room mirror. When it comes to your bottom, a highly informal poll has shown that men care more about what it feels like than what it looks like. And something to grab on to is a good thing.
- **Accidental burps or farts.** The more embarrassed you are, the cuter they're likely to find you. Just don't make a habit of it.
- **Slightly rangy body odor.** I once had a boyfriend who begged me not to shower after going to the gym. Science has proven there are pheromones in your sweat that will draw him closer, so don't go drowning your sex appeal in deodorant and cologne.

muting time? Shacking up is a big decision, and he has a right to think it through. Use this difference in opinion as a jumping-off point to discuss where your relationship is going. If you see it heading in a different direction than he does, a breakup may be in the cards. But don't ditch an otherwise strong connection just to save face.

Someone you're seeing casually has just professed his undying love. You may feel the same way, but you may not. It's too soon!

You don't always want to take a guy like this seriously. He may be dumb/romantic enough to get caught up in the fantasy of love at first sight and all that. Tell him he's sweet, but be a bit skeptical about his intentions. It's not that you aren't completely desirable or anything, but when someone pledges his love for you that soon, he's probably (1) desperate to be in love, (2) genuinely in love, or (3) bananas. It takes time to find out the truth, and if he's truly in love with you, well, then he'll wait for you to decide.

One of you blurts out "I love you" way too early—and the other person doesn't say it back.

If it's you to him: Dang! It slipped out, and all you got back in return was a tepid "Thank you." Was sexual euphoria to blame or were you just being honest? It's best not to overemphasize the awkward moment with apologies or discussions—let him think it through until he is ready to return the sentiment. And don't beat yourself up. Being open and vulnerable is part of being in a relationship.

If it's him to you: On the other hand, if he opens up to you and you don't feel the same way, be kind and gentle. Make sure he knows that you appreciate what he said and that it makes you feel good. If he pressures you to respond in kind, tell him, "I care about you and I take those words seriously. I'm just not ready to say them yet." You can also just dive in for a distractingly passionate kiss—beats a change in subject.

Sticky Note: If you want to know how a man feels about you, watch a sappy old movie together. If he reaches for your hand during the love scene, you're golden.

Why is it that whenever you start having sex with someone, you stop falling in love with him?

This probably means that the challenge of "landing" someone is hotter to you than the actual potential relationship. Slow down on the sex and work on the emotional side of things first. Don't start messing around until you feel a true connection with the person. Once you have that, the sex will just be the cherry on top (sorry, bad pun).

Dear Jane,

Every time I go on a date with a guy or start to casually date someone, my well-meaning friends pry about my sex life. I'm not a prude, but I just don't think it's any of their business. How do you respond appropriately and in a conversation-concluding way without giving anything away?

Replying "none of your business" comes off as defensive (and also suggests that the answer is something that you don't want to admit). Instead, I think this response always works: "Why do you ask?" It forces the questioners to produce a justification for their rudeness. Or you could simply reply: "Would you be interested in buying our video?"

You need to call it off with someone you're dating but you still want to be friends—really.

Tired old lines like "Let's just be friends" or "I value our friendship too much to have a relationship with you" have long been fallbacks for cowards looking to duck out of dating someone. Therefore, in the cases where you actually do want to transition a dude from date to pal, you need to be a bit less clichéd.

Let him know that you really enjoy spending time with him—and cite examples, like your shared sense of humor or mutual love of science museums, before lowering the boom. Tell him that you don't think you are right for each other romantically, and follow up with a platonic invite, like "My friends and I are going to that show we talked about, if you want to join us." You may not get to eat your cake and have it, too—you have, after all, bruised his ego. But if you continue to show genuine interest in him and are careful not to blur the lines (by letting him spring for movie tickets or indulging in a little "harmless" cuddling, for instance), you guys will be Jerry and Elaine in no time.

He has way more money than you and is always paying for things. You want to reciprocate, but you just can't afford it.

Tell him you are going to take him out for a special date, and plan an evening in your price range. Whether you take him out for Thai or to a kiddie arcade, he'll appreciate the effort.

Another option is to offer to pay for the cheaper part of an evening out—he pays for the four-course meal (and two bottles of wine), and you pay for dessert. But it's up to you—if he's loaded, the fact that he's footing the bills most of the time means far less to him than it does to you. It's most likely that he enjoys being the high roller with a pretty lady on his arm.

You'd love to have more sticky situations in love, but frankly, you have no idea where or how to meet guys!

Studies show that meeting guys has less to do with sharing interests than with simply being thrown into proximity with them. Although it's known that similar backgrounds draw people together, being in close proximity does influence the development of our friendships and partnerships.

Adopt the "just say yes" approach. No, I'm not telling you that you should take all comers, but you

You're now officially dating a jet-setter. You have all of two minutes to throw everything you need into a bag.

Once, invited on a romantic minibreak, I breezed through my apartment grabbing just the bare necessities and practically vaulted into my date's convertible, excited to get on the road. I arrived at our hotel four hours later to discover I had brought the toothbrush I use to scrub my grout, a pair of underwear whose elastic was no longer elastic, and the leggings I'd worn to Pilates that very morning. Luckily, we didn't end up going anyplace that necessitated my dressing up. Still, I vowed to never again be unprepared for an impromptu vacay.

So, since you never know when an opportunity may arise, I highly recommend you stash the following in a cute tote and hang it on a hook close to your door so you can jet off at a moment's notice.

should accept all social invitations that come your way. Your friend's office happy hour? Yes. Your college's alumni event in your city? Yes. Your co-worker's holiday party? Yes. Even if these events don't yield a treasure trove of single men, they will help you become more comfortable meeting new people, as well as lead to more invitations. Even a girls' night out could result in a promising fix-up.

If the social events aren't happening, then create them yourself. Throw parties and invite a few wild cards—people you don't know well and would like to know better. Or have each invitee bring his or her own wild card. As the hostess, you have a built-in excuse to talk to everyone there.

Expand your extracurricular activities. Go to museum lectures, join an intramural sports league (kickball, anyone?), or stop by a local bookstore for a reading. A rock show is pretty much a guaranteed man buffet. Practice starting conversations with strangers—ask him if he's seen the next band, or if he's read the novelist's first book of short stories. Still too much pressure? Ease in with someone you aren't too interested in attracting. Soon you'll be confident enough to approach any hottie.

Resist MySpace and Facebook no longer—not only will they reunite you with old acquaintances, they're an easy, informal way to get in touch with someone you meet on one of your new outings.

Basically, you're advertising to an ever-expanding group of people that you are single and awesome. Possible side effects? Making new friends and connections.

Weekend Must-Haves

In a clear plastic zipper bag, stash the following:

- New toothbrush
- Travel-sized tube of toothpaste
- Fragrance sample you like the scent of
- Nail polish remover pad
- Mini tinted moisturizer with sunscreen
- Tube of mascara
- Eye pencil
- Lip gloss
- Convertible lip/cheek color
- Deodorant or antiperspirant
- Three tampons

In a lightweight tote, put:

- Three pairs of panties
- Two bras, one black, one nude
- A pair of black leggings
- A soft, cozy wrap
- A stretchy dress that doesn't wrinkle
- Two lightweight T-shirts or tank tops
- A pair of lightweight pants or a skirt
- A lightweight cardigan, preferably cashmere
- A pair of ballet flats

Throw the zip-top bag in the tote, and you'll be ready to jet off at a moment's notice.

You're nearing forty (or thirty but thinking about forty) and you want to get married and have a baby, and not necessarily in that order . . . no one's calling you desperate, but you hear a certain clock ticking and you feel S-T-U-C-K.

Okay, first of all, chances are that the reason you're single is that you're self-aware, discriminating, and probably have had a lot on your plate. One of the great things about the post-feminist era is that we women have had the right to choose any path we want. We can have great jobs, great travel, pursue our passions . . . but we want to find equally great men to help raise that equally fabulous family. And they aren't always easy to find in those forty-five minutes we have between work and yoga. Choices are wonderful but often overwhelming, and quite often we're not sure whether we're making the right ones. Mr. Sort-of-Wonderful may seem lovely, but we wonder whether there's someone better down the line.

If you're getting to the point in your life where you're really beginning to panic, I offer a few pieces of advice—and remember, there are whole tomes devoted to this subject. First, do not—I repeat, do not—set out *just* to get married. Decide what's important to you. Do you want to have a child? Be with someone? Have a full set of china and crystal in your cupboard? If it's the last, forget the manhunt—throw yourself a big birthday party

Dear Jane,

I'm smart and cute, I have lots of friends, and I'm in good shape—yet I'm almost twenty-five and have never had a boyfriend! I'm not gay—I like guys. They just never seem to be romantically interested in me.

Okay, time for an attitude adjustment. From this moment forward you are not one of the boys, the shoulder to lean on, or the girl who is so easy to talk to. You are no longer playing that role in any man's life. Repeat to yourself: You are girlfriend material.

Now, dress up this future girlfriend self. You know you're cute and in good shape—what makes you feel sexy? Get some date dresses and pull together a cache of wardrobe favorites that accentuate your favorite parts of yourself.

Take a good look at whom you're spending time with. A bunch of platonic dude friends? See above. Couples you've known forever? Sorry, now is not the time to play the fifth wheel. Pulling yourself out of your social rut is a good way to start seeing yourself in a different light.

and tell everyone it's your birthday wedding and you've registered so they can buy you your dream gifts. Trust me, far easier than finding a guy who may turn out to be Mr. Wrong.

If you want a child more than a partner, think about whether you're ready to raise a child on your own and what that would mean to your lifestyle. That's no more nightly trips to the gym, no more sleeping in on Sundays. Do you want a baby because everyone else has one or because you think you should? Studies have shown that children don't actually make your life happier. I would suggest they make your life fuller, but if you want this for any reason other than desperately wanting to care for a child (not *dress up a child*), then don't do it.

But if you truly long to fall in love and settle down, then let's think about that. Has something been holding you back? Look at the past few years. Has there been a pattern to your relationships? Have you even made time to have a relationship? Are you waiting for Mr. Perfect to swan in, à la *Pretty Woman*? Because let's not forget: She was actually a hooker!

It's time to be realistic. To have a great relationship, you need to work at it! Work harder at it than you have at any job. That does *not* mean you need to do everything for your man, but it *does* mean you should look at things from his perspective. Does he have a hobby you might enjoy trying out? If you aren't really a cook, maybe now's the time to give it a shot—for example, with the

Now, start flirting! Consider putting up an online dating profile, if only to define your new identity to yourself. Make no apologies for your lack of experience; just state exactly what you know to be true—you are smart, hot, and popular. You can expect a barrage of responses, so try a few practice dates and get your patter down. Enjoy a guy's eyes lighting up when he sees you, even if he may not be the man for you. Another benefit to online dating? You can limit your search to men who are interested in relationships. There's less of the guesswork that goes into meeting a man out on the town and not knowing whether he is interested in a hookup or something more permanent.

The point is to stop thinking of yourself as the girl who never gets the guy. To quote the late, great Funkadelic, "Free your mind . . . and your ass will follow."

step-by-step man-friendly recipes on thepioneer woman.com. (With millions of readers, four children, and a happy marriage, she must be doing something right!)

Other practical strategies:

- Go beyond your traditional circle of people to find men. I know you've heard this advice, but you *have to do it.* Men do not rain from the sky (typically).

- Do some soul searching about what you want in a man, as opposed to what you *think* you should want in a man . . . because they're often two different things. It's not about meeting your best friend's or your mom's expectations—it's about meeting yours.

- Put your best foot, and face, forward. You may expect this date to be a dud, but always look your best. That doesn't mean layers of makeup, but even at the Laundromat you may meet that certain someone, and looking a little spiffy can give you the confidence you need to ask to borrow a quarter.

- Go out with the guys, not always the girls! If you go out with a pack of girls, you might lose the chance to connect with *your* Mr. Wonderful. Let your best guy friends go out on the hunt with you.

- Once you find someone, don't overthink it, and don't assume *this is the one!* Nothing is worse for a man than feeling as if he has to ask for your hand by the end of date three. Relax, have fun, and give your relationship time to see if it works.

- Don't break up for dumb reasons. So he leaves wet towels on the floor. So he's two inches shorter than you. *Come on!!!!* You're looking for a partner, not a custom-made robot. No one is perfect, my love, not even you.

- Hold on to your best assets. Look, I'm not a prude, and I'm not telling you to play games. Just wait until you decide whether this relationship is really a keeper. You'll cause yourself less grief and give yourself a better chance to let the relationship develop in the right way.

- Trust the experts. Go get yourself a subscription to *Cosmopolitan* magazine. Look, there's a reason two million women read it every month. They've got great tips, info, and advice. Trust me, your man will appreciate it.

- If you want to change your body or your lifestyle, look to the Web for an endless cornucopia of support, from diet and exercise plans to message boards populated by eager experts with too much time on their hands. And—cattiness aside—when you need a boost, or just a shoulder to cry on, isn't someone with too much time on her hands a good thing?

Finally, remember, life is a journey, and it is yours to enjoy. We all want a partner, but there are many people I know who I think might have been better off alone. Take care of yourself, and your needs, and the rest will follow.

Online Dating

You feel like a total freak. Even—okay, especially—online.

Don't be afraid to let your freak flag fly! There's a niche dating site for every type of person. Unlike general sites such as Match.com that cast as wide a net as possible, niche online dating forums are self-selecting, based on interests, ethnic makeup, or dating preferences. Examples include MulletPassions.com, LargePassions.com, GolfMates.com, DeafSinglesConnection.com, SingleParentLove.com, and BikerKiss.com. Got a quirk, hobby, or passion? Find a like-minded man through one of these sites.

The biggies—Match.com, eHarmony.com, Chemistry.com, Nerve.com—have the benefit of thousands of users to choose from. Narrow down the seemingly endless options by searching by location, age, or interests—even by height or weight (not that you'd ever be so shallow).

Be sure to take some time to hone your profile. If you're not a writer, recruit a friend to help you fine-tune your ad. Make the profile memorable by mentioning the things that make you unique: reveal a guilty pleasure or an embarrassing moment from childhood. These things can often serve as talking points when a man sends you a message.

Pictures are important. Your profile is far more likely to be looked at if it comes with a photograph. Resist the temptation to get your graphic designer friend to airbrush the hell out of it. You always want to represent yourself at the moment you are placing the ad, even if you secretly want to lose weight or change your hair color. A clear photograph of a friendly, smiling person is always going to do well.

You're trying to be hip when communicating online, but between the strange symbols and mixed messages, this new language is far more confusing than just waiting for the phone to ring. You just got a ;) and you're not quite sure how to take it.

You've been winked, at my friend, and like in the nondigital world, winking is a fairly risk-free way of showing interest in a person. It also doesn't mean a whole lot. If you're interested in the winker's profile, wink back and see if he responds with an actual message. Or message him yourself: "Thanks for the wink—what's your story?"

You've been messaging back and forth, but no date. What's up?

Foreplay is great, but we ultimately all want to get to the main event, right? This love letter purgatory could be the result of simple shyness or insecurity—many people have an easier time communicating through writing than through speech. The medium allows a person not only to think through his or her responses but also to fit flirting into a crammed schedule. In this case, you may want to give the guy a nudge. Suggest that you meet in person to further discuss your mutual love of experimental fiction.

It's also possible that the dude is just enjoying the flirtation but, for whatever reason, has no intention of meeting up. This scenario is a good reason not to indulge in prolonged e-mails before a first date. Why waste your time? Another good reason to get the party started: two people who have top-notch banter may also have zero chemistry. All it takes is a quick drink to discover where you stand with someone.

You've met the perfect guy online . . . but he lives across the country. Should you buy a plane ticket?

Start with Skype—it's free and will give you the chance to see your man in motion. Still interested? As romantic as it sounds to run away for a weekend of passion, it is best to build a meeting into a bigger trip. Have a friend you've been wanting to visit in his hometown, or a business trip a state away? In the case of a risky meeting like this, you want to have a clear escape route (and an alternative fun time) in the plans if your attraction turns out to be limited to pixels.

Your long-distance guy turns out to be as irresistible in person as he is in e-mail.

Prepare yourself for some steamy weekends—and some big decisions down the road. A long-distance relationship can keep the infatuation fresh, but it can also leave you feeling lonely and unsupported on the day-to-day stuff.

You think this guy's a little *too* perfect. He must be married, right?

Okay, there are some bastards out there who use online dating to get a self-esteem boost or to find their next (often unwitting) mistress. A little judicious Googling now is okay when looking for previous—or current—marriages! Otherwise, look for the usual signs—is he always unavailable on weekends, or does he never invite you over to his place? Trust your instincts.

You receive a message from a man you don't find physically attractive.

It's sometimes hard to comprehend that your interactions with others on online dating sites involve real people, so it's always best to react the same way you would in a real-life situation. Think about it: all day and all night, men make advances to women—on the bus, on the train, in traffic, in elevators, at the grocery store, and, of course, in bars. So, just like in any other situation where you are approached by a man, respond politely. If the messages continue, how about a simple "No thanks" or "I don't think we're a good match, but thank you for messaging me"? Let the guy down kindly and promptly. For the most part, he will accept your lack of interest as part of the risk of online dating. Of course, there are the notorious e-mails littering the interweb written by men unwilling to accept rejection. If you get a nasty response, you have every right to ignore his message or block him (or maybe even forward his message on to Jezebel.com).

You've messaged a guy who looks perfect for you and he hasn't responded—what gives?

Sadly, this is all part of the risk of online dating. You can't know why he's not interested—maybe he's just online to browse and not to date, maybe you look like his nasty ex-girlfriend, maybe he started dating his coworker the very day you sent your message. Or maybe he doesn't think you are a good match. It's best to forget him and find a new online crush.

There's no chemistry on a once-promising blind or Internet date.

This is why it's best to meet up with an interesting Internet swain before investing too much time in phone calls or letter writing. Sometimes a great pen pal just doesn't add up to a hot date. Cut the evening short and follow up with a simple "I'm sorry, but I just didn't feel we had a connection." It's all part of dating.

Dear Jane,

A few months ago, I met my boyfriend on JDate, a dating site for Jewish singles. He assumed I was Jewish, and I didn't correct him when I had the chance. Now I'm afraid that if I tell him, he's going to dump me. What do I do? I really like this guy, but it's getting weird.

Presumably you haven't stooped to sharing Passover "memories" or faking bat mitzvah photos. So the next time you see him, give him a kiss, tell him you have something to confess, and let him have it—you're a shiksa. Tell him you never meant to deceive him, but just got caught up in the thrill of meeting someone you really, really like . . . and did you mention you really, really like him? It may be the case that he does only date Jewish girls and your confession could lead to a bigger discussion. Often a non-Orthodox Jewish guy just wants to know that his children will also be Jewish, a distinction that is passed down through the mother. If you're willing to convert (and presumably you are if you were on JDate), this may be the time to let him know. Most likely, the deception is going to be more upsetting to him than the cultural difference. Prepare yourself for some rebuilding of trust.

Where it's okay to exaggerate on your online profile:

Okay	Not Okay
Five-month-old photo	Five-year-old photo
Talking about your job	Bragging—or lying—about your job
Using an alias	Impersonating someone else
Listing your favorite quotations	Pretending you said them
Photo of you enjoying a cocktail	Photo of you doing a kegstand
Hot beach photo	Hot hotel-room photo
Not posting your age	Lying about your age
Posting a photo of your pet	Using your pet as your profile pic
Saying who your fave band is	Bragging about sleeping with them
Updating your status	Not updating your marital status
Saying you love cop movies	Saying you love handcuffs

The Ex Files

You just ran into your ex with his new girlfriend and suddenly you can't stop thinking about him.

We always want what we can't have. The same ex-boyfriend who was so irritating last week suddenly seems appealing when he's on the arm of a new girl. It's okay to savor a few fond memories as long as it doesn't lead to any drunken texts or dramatic scenes. The feelings you had for him before seeing him at brunch with that blonde are most likely to be the accurate ones. So give yourself a few hours of delicious nostalgia, then refer to the laundry list of his annoying traits (write one up, if need be) and move on. And find a new place to have brunch.

If you catch sight of your ex with his new lady, suck it up and walk over with confidence. Even if you're slouching around in track pants on your way to the Laundromat (and isn't that always the case?), smile big and shake hands. The fact that you can be confident with your singlehood (or at least fake it) is inherently cool. And, if need be, it's fine to point out the awkwardness of the situation, as long as you do it with a sense of humor.

You run into an ex while with your current flame.

If you're the one with the new hottie on your arm, try not to rub it in. Be polite, introduce them, and, if possible, be on your way. It never hurts to be classy, even if you were the one who got your heart broken. Seeing you with a new man is punishment enough without forcing your ex to witness a PDA.

Dear Jane,
To get revenge after my boyfriend broke up with me, I started seeing his best friend. Now the friend wants to get serious. How do I tell him I was just using him?

Well, sorry, but you've both behaved like assholes—you shouldn't have used him to get back at your ex, and he shouldn't have hooked up with his best friend's ex. No need to make the situation worse with overt honesty. Just tell him you aren't ready for a relationship and stop seeing him. He took the risk of being a rebound (and of losing his friend), and those are the consequences. And next time don't involve someone else's feelings in your revenge fantasies.

You and your ex have actually managed a great friendship, but so far you've avoided the subject of new romances. Now he wants you to meet his new love, and you're not really feeling it.

If you want to embark on a true friendship with your former man, you're going to have to accept all aspects of his current life—including Miss Right Now. That doesn't mean you have to go to dim sum with the two of them every Saturday, but you should aim to meet her in a neutral, low-stress scenario. Plan to meet up at a mutual friend's party (bring backup if necessary). Chances are you'll probably like her—he always did have good taste in women. If you truly can't handle it, you should probably rethink your "friendship." Do you really want to be friends, or are you waiting around hoping for more? If it's the latter, graciously suggest you two take a break for a little bit.

You want to date your friend's ex.

This is an area in which to tread lightly. Always ask your friend's permission before dating her ex, even if the ex in question is a guy she made out with ten years ago at a rock show. And respect her response, even if you consider it illogical. If you decide you can't live without the love of her college ex-boyfriend and she wants you to keep your mitts off, you will quite possibly sacrifice your friendship if you go ahead with the guy.

Likewise, if she wants to date your ex, you're justified in being upset if she doesn't consider your feelings first. If the dude in question is a recent ex, someone she knew only as your man, you're justified in thinking she's bananas (and a bad friend to boot) for making the request. However, if she asks for your blessing to date an ex you no longer have feelings for, even if it gives you a pang of possessiveness, you may want to relent if you think the two crazy kids could make a go of it. It's just good man karma.

Your boyfriend's ex wants to be BFFs with him.

It's natural to want your man's ex to be firmly out of the picture, but try to keep things in perspective. Yes, they have shared experiences that lend themselves to friendship, but remember that there's a reason they're not still together. Don't be

Dear Jane,

My ex-boyfriend recently posted one of my gooey love notes on his blog. I'd rather the world didn't know that I used to refer to him as "Big Daddy." Can I compel him to delete the letter?

It doesn't hurt to ask. Just try not to call him "Big Daddy"—or "Megalomaniacal Asshole"—in your e-mail. Say that the letter was a private exchange during a private relationship and that you'd prefer that it remained between the two of you. Technically, although the letter itself is legally his, the copyright to the writing ("sugartits" and all) is yours. If it comes to it, you could threaten to sue. Just be careful that this doesn't result in even more attention being drawn to your missive. Most likely, if you just sit tight, the whole thing will blow over before you know it—it's not like anyone reads his stupid blog anyway, right?

concerned if he has a casual drink with an ex or if mutual friends bring you into the same social circle. On the other hand, if they have unresolved issues or argue about their former relationship, put your foot down. You're not an extra in their traveling show. Suggest that they spend some time apart from each other until the emotions cool.

You have a stalker or obsessive ex.

First, tell him in no uncertain terms to leave you alone. If he persists, start keeping track of every call and visit and save voice messages from him. Tell security in your apartment building and at your office that you're being stalked and (if possible) give them a photograph of him. This is especially important if he is a recent ex whom they were used to waving through the door.

To get a restraining order you need to have proof that you are in physical danger from your stalker. The court will likely ask you specific questions about your interactions with him—this is where your records come in handy. Once you've been issued a restraining order, keep a copy of it on you at all times and call the police if he violates the decree. In the meantime, have friends and family members walk you to your car at night and let your neighbors know what you are going through.

In Bed

You haven't had sex in a really, really long time, and can't decide whether to break the seal or wait for love.

Consider why you've been holding off in the first place. Is it because you've been busy launching a new business or finishing your dissertation? If you've simply been concentrating on other aspects of your life, you may want to jump-start your return to dating with a (safe) fling. On the other hand, if the reason you've been staying out of the action is because you equate sex with a relationship, a down-and-dirty affair is unlikely to satisfy you. Stop torturing yourself with an arbitrary timeline and work on the dating thing. And don't worry about your inevitable return to the land of the sexual—it's just like riding a bicycle.

You've been dating for a few months when something starts to feel itchy. Is now the right time to talk about STDs?

This question is always a tough one—you don't want to bring it up while you're still making out on the couch, but if you wait until he's reaching for the condom, can you even expect honesty from a man with an erection? Wait until the pants are coming off and then say, "I just got tested for STDs and I'm a clean teen [this had better be true, BTW]—how about you?" Volunteering the information makes it clear you don't consider him just a dirty man ho. But no matter what the moment is or the response you get, you are always taking a risk. People lie about having been tested and they lie about having diseases. Use condoms until you are in a relationship, and get tested regularly.

You have an STD. How and when do you 'fess up to your bed buddy?

This is definitely a question of when, not whether—you *must* tell your prospective sexual partners if you have a communicable disease. Never risk another person's health because "it's so unlikely that I'll pass it on." Tell your date in a neutral setting that's not sexually charged (i.e., don't wait until you are grappling on his kitchen floor). Calmly relay the facts of your disease—how it is transmitted and how likely it is that you will pass it on during safe sex. Then give him time to think things over, and be prepared that he might bail. Just know that, no matter what you have, you're in good company. There are plenty of dating sites and support groups for people with STDs if you want to find a partner who knows exactly what you are going through.

You get caught with your pants down—literally.

If you are having sex in a shared space, well, pull up your pants and apologize. If kids are involved, you might have to give the old "when two adults love each other very much . . ." speech, which is no fun for anyone. Or if they are young enough, do what my friends do and tell then you were wrestling. Though then, of course, there's the danger that they'll want to join in! If you've been caught by someone over drinking age, just promise to be more respectful in the future and change the subject. Chances are he or she is as embarrassed as you are.

Now, if the Peeping Tom or Thomasina entered your home or bedroom without permission, you shouldn't be the one apologizing. Have a discussion with your children about respecting privacy, and tell roommates and friends not to come a-knockin' when the futon is a-rockin'. In the case of your parents, seeing their kid in flagrante was probably punishment enough.

Your mate has erectile dysfunction.

In this situation, you need to be understanding and he needs to get you off. Suggest that he see a doctor or a counselor to find out whether the problem is physical or psychological and to look into

Dear Jane,

If I'm hooking up with a guy when I'm having my period, how do I let him know without it being awkward (and still getting a little bit of action)? Is there an etiquette protocol for this one, and is it safe? What are the cons of doing this?

Awkwardness (and a little mess) may be unavoidable, but it would be hard to find a grown man who hasn't played around with a woman having her period. Tell him straight out in the makeout phase of things: "So, it's an inconvenient time of the month for this, but I'd still like to get busy." Suggest continuing things in the shower, or throw a towel on the bed. You can also hold back your flow using a diaphragm or an Instead Softcup.

Sex during your period is perfectly healthy, but don't forgo the condoms. It is still possible to get pregnant while on the rag, and women are more susceptible to STDs because the cervix opens up to allow blood to pass through, creating an entranceway for dangerous bacteria.

options like Viagra or therapy. But, more important, accept him as he is and let him know that you can have pleasurable sexual experiences without his hard-on. Experiment with toys and massage, or just make out like teenagers. By taking some of the pressure off his erection (so to speak), you can relieve some of his performance anxiety, which might even solve the problem. Just be patient . . . and creative.

Dear Jane,
My boyfriend likes to go out with his buddies and drink—something I am totally fine with. The only problem is that when he gets home, he's drunk, wants to shag, and proceeds to have sloppy sex with me, if not whiskey dick, which ruins it altogether. I love the guy, but it's annoying and happens often.

Talk to him about the problem while he is sober. Say, "Listen, love, I want you much more when you aren't doused in drink. You can get lit with your dudes or you can get it on with me, but you can't do both the same night." Make sure he understands that this is not an ultimatum. He's free to hang out with his friends and imbibe, but he has a responsibility to you as a sexual partner to have mutually satisfying nookie. This can't happen when he's tanked and has a softie.

His penis is too big—or too small.

Supposedly a huge "member" is every woman's fantasy, but in reality, sleeping with someone with a too-large penis can be uncomfortable and even painful. At least discussing this problem won't harm his ego. Tell him he has a monster cock—the biggest you've ever seen—and suggest that you use lube or extra foreplay to make sex more comfortable for you.

Got yourself a short man instead? Thankfully, the saying "It's not the size of the wave, it's the motion in the ocean" happens to be true. Experiment with rolling your hips or having him rock his pelvis back and forth when he's with you. The extra effort will up your sensation tremendously.

Your mate wants to have a threesome.

Well, are you interested? If so, it should absolutely be on your terms. You get to choose the girl (or guy) who joins you and you get to decide how much action either of you participates in. You and your man should set guidelines—are you comfortable with him penetrating another woman, or would you prefer him to be voyeur to the girl-on-girl? When it comes to selecting a third party, skip your friends and peruse the online dirty personals or find an old acquaintance—preferably from out of town, whom you won't see often and who is less likely to spill the beans to mutual friends. As in any Internet dating situation, you want to do some careful vetting—have drinks with your prospective double date and work up to the main event. But be very, very careful—this could really change the dynamic of your relationship, so think hard before you leap in this regard. And here's the thing: If you aren't okay with the word potentially getting out that you had a threesome, don't do it! If there is one thing a man likes more than a threesome, it's telling *other* men about his threesome.

If you aren't interested, let him know and try to be supportive of his fantasy. Explore the idea with dirty talk or watch some three-way porn together.

Dear Jane,
The guy I am seeing is otherwise totally normal—good job, good family, good friends—but he has, ahem, weird sexual requests and fetishes.

When it comes to kinks, sex columnist Dan Savage has popularized the three G's: "'Good, giving, and game' is what we should all strive to be for our sex partners, as in 'good in bed,' 'giving equal time and equal pleasure,' and 'game for anything—within reason.'" Words to live by, as long as you remember the "within reason" part. You should never feel forced to do anything that disgusts you or endangers your health, and your partner should be understanding of that. If he's selfish or forceful, then you should consider ditching the relationship, no matter how eligible he is in other ways. However, if he's patient and careful and you don't mind his fetishes (maybe you just think they're weird), then why not try them out? Being covered in baked beans (sploshing), having your toes sucked (shrimping), or being rubbed against

You fight every time you have sex.

Is arguing an aphrodisiac for you, or are you just fighting about bad sex? If you need to be charged up to get it on, you might be substituting one kind of passion for another. Simply discussing this propensity will help you avoid the habit of picking fights. Instead, try incorporating some role play that will allow you to have some hot arguments without damaging your relationship.

If bad sex is your problem, try to take away the pressure of a climax by slowing things down. Play at being inexperienced teenagers again—spend a night holding each other, work up to removing clothes and rounding the bases. Consider holding off on full intercourse for as long as it takes for you to rebuild your intimacy.

He wants sex much more frequently than you do—or vice versa.

There are so many ways to deal with the truly common problem of differing sex drives within a relationship. Usually one would start with an adult discussion that, ideally, concludes with the two parties agreeing to work on it. Does he like to have sex in the morning and you like it at night? Try to compromise with your schedules—morning sex on weekdays and night sex on weekends, or vice versa. Not every sexual encounter needs to be full-blown—have a quickie, or try mutual masturbation. Finally, on occasion, the one of you with the stronger drive is just going to have to take care of him- or herself. The key is incorporating these solo trips into a fuller, more flexible sexual relationship.

(frotteurism) may not turn you on, but if it gives him pleasure, what's the harm? And while you are indulging his fantasies, you might discover a few of your own!

He wants to videotape or take photos of you.

I'm no prude, but let me say two words of caution: *the Internet*! Whatever his intent, or yours, just think about how you would feel if somehow your private video or photos got on the Internet and your parents or your future children saw it. I could tell you about countless sticky situations—the daughter who forgot she took pictures with her boyfriend and gave the camera back to her mom, a teacher who then brought the camera into class. Or the daughter who caught a glimpse of Mom with her new boyfriend on a wild weekend away. Ewwwww. So if you want to star in your own private porno, that's your business. But use *your* camera and destroy all the evidence *that night*!

It's Serious

Your new boyfriend won't give you any alone time. (When are you supposed to shave your legs?)

Suggest a hot date for a few days in the future, assure him that you can't wait, but make it clear that you can't see him until then. If he protests, give him a kiss and tell him you need a little time to do laundry and make yourself beautiful. (Not to mention poop.)

Your new boyfriend has become entirely too close to your mom.

Well, at least they don't hate each other, right? Most likely they're both invested in the relationship because of you. If she's getting too nurturing (cooking him dinner, doing his laundry), have a talk with her about giving you two some alone time. And tell him that you're a grown-up and, as awesome as she is, you can't see your mommy all the time.

Your parents hate your new boyfriend.

Now, what if the opposite is true and your mom (or dad) thinks your new Mr. Right is just another in a long line of Mr. Wrongs? Calmly explain that what makes you happy is different from what makes them happy, and try to point out your boyfriend's best features (other than those between the sheets). But *Father Knows Best* wasn't just a show on TV. Take a moment to consider whether your parents might be right. If, in fact, your choice in men leans toward the unemployed, unfriendly, parents'-worst-nightmare type, you have to ask yourself whether you're really in love or just shacking up with Mr. Inappropriate in order to drive your folks a little crazy. If the latter, scrap the chip on your shoulder. Bad men are easy to find; good men aren't. Quit wasting your time, and save your poor parents some heartache.

Your boyfriend and your best friend don't get along.

Is the problem a personality clash, or is your boyfriend (or friend) jealous of your close relationship? If it's a jealousy thing, be warned—it's a controlling man who tries to separate you from the people who make you happy. But if the two of them simply can't

Dear Jane,

My BFF has a new boyfriend, and whenever we're all together, they can't keep their hands off each other. How can I tell them their behavior makes me uncomfortable?

It's best to talk to your friend about the situation when her man is not around. Tell her, "I'm happy you're getting some action, but the PDA thing is more than I can handle. Can you guys stay on first base while I'm hanging out with you?" It's her responsibility to talk to her boyfriend and tone it down. And if they still attempt to round second or third while you are with them, point it out in a lighthearted fashion. Say, "Is this your way of recruiting me for a threesome?"

see eye to eye, spend time with them separately, and be sure to reassure your girlfriend that she won't suddenly be banished to an unsatisfying relationship with your voice mail. You don't all need to hang out in one big group. Just have them agree that for special occasions, like your birthday, the two will call a truce for your benefit.

How about if you hate your boyfriend's bro-friend? If he's rude or insulting to you or disrespectful of your relationship, don't take any crap. Tell your boyfriend exactly why you don't care for his pal and give specific examples. Sometimes a guy will keep an old friend around for the same reason he's been wearing the same underwear for the past ten years—simple laziness. If you stay calm and don't issue any ultimatums, your boyfriend might come to the same conclusion you have and start the separation process. If, on the other hand, you can't stand your man's friend simply because he's uncool, awkward, or in a different place in his life than your boyfriend, just grin and bear it. It speaks well for your mate that he doesn't ditch people from his past.

You find out your best friend's mate is cheating on her.

Don't you wish you could travel back in time for a moment and cover your ears? This is such a painful predicament. As uncomfortable as it is, you must confront the philanderer and tell him that if he doesn't tell your friend, you will. Give him a set window of time (I think forty-eight to seventy-two hours is fair), and check in on him. Keep your fingers crossed that he does what he's told. If you do have to tell your friend yourself, be prepared that she might not believe you. She might also lash out. Don't take this personally. Every time you get frustrated with her, just think: This could be you one day. How would you want to be treated? There's one exception to this rule. If you truly, truly know (and I mean you're 100 percent sure) that *she* knows he's cheating, don't confront him. If for some reason she's chosen to turn the other cheek on this issue, perhaps you should be talking to her about her own state of happiness, not fidelity, to see if you can help.

Your mate talks smack about your parents.

We're all free to bitch about our parents as much as we want, but woe betide the pal who chimes in to agree. Your man needs to be respectful of your family, especially if he wants to stick around for a while. Consider the remarks he's made. Is there any truth to them? Do they stem from the way your family treats your boyfriend? If your relatives are partially at fault, let him know that you realize that your family can be difficult, but explain how awful you feel when you hear your parents being insulted. Ask him to be more respectful of your feelings and, while you're at it, talk to your family about how they treat him. If his criticisms are out of left

Dear Jane,
My boyfriend doesn't want me to do anything without him—so, no more girl time. But he won't include me in plans with his friends, either. So I'm either handcuffed to him or alone.

Listen, this just doesn't sound like a healthy relationship to me. I love undivided attention as much as the next girl, but if he's too afraid to give you time with your friends, there's something deeper going on here. First, assure your man that you love him, but that you also need to watch *American Pie* and eat Ben & Jerry's with your girls. If he's always monitoring how you spend your time, you need to consider whether he is too possessive or even abusive. He should want you to have a support network on your side (as well as someone besides him to complain to about soft-cup bras).

If he never includes you in plans with his friends, let him know that you want to be part of all aspects of his life and ask him to loop you in on an outing. Make sure he knows that you are happy for him to have alone time with his boys but that

field or, worse, have to do with things like their socioeconomic class or education level, you may want to think about whether he's trying to divide you from your family. Abusive assholes are good at driving wedges between a girl and her nearest and dearest, so be forewarned.

Your boyfriend does everything he can to keep you away from his family.

Often, his decision not to introduce you to his family doesn't have much to do with you personally. He might find his family difficult or embarrassing and, in order to keep life simple, chooses to keep church separate from state. But if you've been dating for a significant amount of time, it's natural that you should expect to meet his family, if only to know that he takes you seriously. Tell him exactly why you want to meet them, assure him that you will be accepting of any quirks they may have, and explain that it's important to you to be part of every aspect of his life.

You've convinced your boyfriend that meeting your family could be dangerous for his health.

If you're the one keeping your kin under lock and key, you might want to consider why you're doing so. Is it just because your annoying aunt will start sending out wedding invitations if she catches sight of a man on your arm? Or are you ashamed of them? You should know that if you plan on sticking it out with this guy, he's going to have to accept your family as they are, weird food and all. Tell him about your fears, bite the bullet, and introduce them. It's likely that he won't even notice all the little quirks that seem so noticeable to you. And who among us doesn't have a few strange family members?

If you're ashamed of your boyfriend, you might want to reconsider the relationship. Why are you with someone you don't consider good enough to meet your parents? You can't change your family, but you can change your man.

You have a huge fight with your boyfriend's sibling or your boyfriend and your brother are acting like cavemen.

As they say on *Arrested Development*, family first. A relationship with a brother or sister is everlasting, whether we like it or not. You have to make peace with your boyfriend's nutjob sister, if only for his benefit. If she is truly an awful person (as people's siblings sometimes are), try to keep your interactions with her limited to birthdays and holidays.

If your boyfriend and your brother don't get along, have a chat with each of them individually. Ask them to play nice on your behalf and appeal to their masculine sense of honor. They'll be on grudgingly good terms by the next Sunday dinner.

you also want to be included on occasion. Most likely, he's just trying to spare you too many evenings of bad beer and Rock Band.

Dear Jane,

My husband's annoying habits—the sounds he makes when he eats, the way he talks to me when he's in the bathroom—are extinguishing any sexual desire I have for him. How can I tell him this without seeming cruel?

In every long-term relationship, we naturally start to hate each other's quirks, if solely through overexposure. Be forewarned that even if your husband were to stop chewing with his mouth open, you'd likely find some other behavior infuriating. Without nagging, work on setting some boundaries—like no toilet-side conversations. Make sure you're taking time every day for yourself, whether through a class or just a long shower with the radio on. A little private time can be a major aphrodisiac.

If you can—and believe me, there are days when this seems impossible—for every little thing that drives you crazy, find one great thing he does. If you always focus on the negatives of a relationship, your relationship will be defined by what's wrong with it. If you choose to focus on what's right, the wrong thing won't go away, but it will seem much less important.

Your mate loses his job.

The first thing to do is to be as calm and supportive as possible. He knows that he needs a job, that the market is tough, that you have bills to pay. Take some time to let him know that you love him for more than just his moneymaking potential. Use some of his free time to have more sex (it won't break the bank account, after all). And when he's ready, help him take all the necessary steps toward finding a job.

Your boyfriend and your boss have become chums.

Tell your man that work needs to be separate from your private life. You spend the majority of your life at work, and you don't want to come home at night to find your boss on the couch playing video games. Ask your boyfriend to keep his bromance limited to office events.

Your boyfriend feels entitled to comment on your body and what you should do to perfect it.

A resounding f—— you to this guy! If this isn't a question of health, just aesthetics, he's being a controlling prick. Tell him you need to feel that he accepts you as you are or you can't be with him.

Someone important to you doesn't support you the way you would like.

Not to state the obvious, but death is very painful. And unless you've experienced the death of someone close to you, it's truly hard to know what such a loss feels like. As someone said to me when my mother died, "You've just joined a club you never wanted to be a part of." She was fifty and I was twenty-one, but I felt closer to her than to anyone my own age. And no one really knows how to deal with someone who has had someone close die, since we all need very different things when we're in mourning. In fact, you'll probably be so overwhelmed with the loss that you have no idea what you actually want in terms of support. Make it clear that you need his shoulder to cry on, but don't feel that you need to justify to him the way you show your emotions. Likewise, don't assume that his stoicism means he doesn't care. The death you're dealing with might have brought forth his own fears of mortality. Or he may just not be capable of emoting in the same way. Ask him for the help you need, but don't worry if he can't share in your mourning.

He never does anything for your birthday.

Welcome to my world—and that of my poor husband. If he does nothing, he loses. If he goes all out, he loses, too—because unfailingly he waits until the last minute to make plans, and ends up spending tons of money on something haphazard. Princess problems, right? I used to take all of this really personally—where's my cheeky engraved stationery? My gift certificate to the newest eco-spa? My French macaroons? But I've come to realize that I often set my husband up for failure—if I don't really know what I want for my birthday, how is he supposed to?

It's not fair to assume your man is a mind reader. Let him know that it's important for you to be celebrated on your special day, and share whatever you're secretly wishing for, no matter how major, insignificant, or bizarre. No fair answering with "I don't know; what do you want to do?" He can offer you a few suggestions or you can tell him exactly what kind of celebration you want—a dinner for two, a scavenger hunt, a "surprise" party? Set the date and draw him a map, if necessary, but the key is to leave him in charge. Once the big day unfolds, let him know exactly how grateful you are for it. It shouldn't be such a tough sell next year.

The fire is out—you're stuck in a rut.

The reason fairy tales end at the wedding is that no one wants to see the happy couple five years into the relationship, when she's a bored hausfrau and he's schtupping the handmaiden. Relationships take work, especially when it comes to trying to get back some of that obsessive infatuation that came so easily in the beginning stages of your union. Start with yourself. What makes you feel sexy? Toss the worn-out sweatpants and stretched-out T-shirts you sleep in and get a soft little nightie. Get your hair done and try on some new colors at Sephora. Do what it takes to get you back to the place where you're checking yourself out in store windows and your rearview mirror. Once you have your hotness back, start flirting. Grab your man's ass in the kitchen, compliment his baby blues, or wink at him over the dinner table. If you start bringing back the behaviors that marked the beginning of your relationship, you'll start reigniting some of that fire.

You can't stop fighting in front of your children.

The two of you need to make a concerted effort to move the arguments behind the scenes. Your children need to feel as if they have a stable home life, even when their parents are disagreeing with each other. Designate a space where the two of you can have the privacy to work things out, and then when you start to argue, take a deep breath, excuse yourselves, and resume the conversation in your bedroom or study. Consider speaking with a counselor about calmer and more productive ways to discuss your differences.

You need to tell your partner you've been unfaithful.

First examine why you strayed. Were you just craving a crush? Were you bored or drunk? Or were you trying to get out of your relationship? If you know you've made a mistake and want to stay with your man, then you're going to have to earn his trust back. It's best to tell him as honestly and simply as possible that you cheated on him and prepare for some tough times ahead.

If he's cheated on you, you need to decide whether or not it's a personal deal-breaker. If there's cheating early on in the relationship, it can be hard to know if the guy is just getting something out of his system or if he's a jerk. That's the thing about dating: At the end of the day, you're having intimate experiences with someone you don't know very well. You don't know his personal code of honor; you don't have his dating rap sheet. You have to go by instinct, how much you care about him, and what he has to say for himself about it. Don't date any man who doesn't know why he does things. Talk things through and let him know that it'll be a slow road back to regaining your trust. And, whatever his excuse, don't ever let any man blame you for his infidelity. No matter what your relationship problems are, it's unacceptable that he should step out on you instead of confronting the issues.

You're done. It's over. You want out of the marriage.

If you've done the hard stuff—the long discussions and the marriage counseling—and just can't make it work, then it might be time to talk divorce. Both you and your spouse will have to get lawyers, but you might also consider hiring a neutral mediator to help you divide assets and debt in a calm and civil way. Be prepared to sacrifice some material things for your freedom—no toaster oven is worth a bitter argument. And don't forget that a marriage is bigger than the two of you—even if you don't have children, you have friends and family who have had a long relationship with your husband. A breakup can be like an earthquake, with ongoing aftershocks. Be prepared for him to find a girlfriend. It will happen sooner or later. Be prepared to be talked about. That will also happen. This is going to be a long haul.

If you have children, you must be as unselfish as possible. This divorce needs to be at least as much about their needs as it is about your own. While I'm no expert, I strongly suggest you seek out a child psychologist well before you make the decision to physically separate. While divorce is difficult on children, there are ways to make it easier, starting with reassuring them that your feelings about them haven't changed and that you'll do your best to spare them as much misery as possible.

PART 2

Sticky Situations in Weddings

I once attended a wedding where the bride was so nervous she broke out into horrible hives, and the tailor had to bring in extra fabric to make a last-minute wrap to cover up her red, irritated chest.

While my skin, thankfully, behaved at my own wedding, my ceremony had its share of snafus. After planning my perfect wedding to the nth degree, I left the rings at home. Everybody switched the placc cards after I'd spent hours calculating my dream tablc combos, because I couldn't afford a wedding planner to monitor guest activity. And then my seven-year-old cousin decided to play "Free Bird" just as the music for our first dance was meant to commence. But, looking back, I realize my woes were pretty minor.

At a rehearsal dinner I attended years ago, an old high school friend of the bride toasted her by going on and on about how much she loved her, "but not in a lesbian way." Fear washed over the bride's face, and her grandmother's jaw dropped to the floor. But this was just the beginning. The friend continued on to say that the bride had slept with the majority of the groomsmen. And she wouldn't shut up until she was literally pulled off the stage. From this moment on, I knew never to drink before giving a toast—and that in this case, shorter is definitely sweeter.

Because there's so much to do at a wedding, so much can go wrong. And because of the obscene level of importance with which we endow the day—trust me, we'd all be better off if we took our marriages as seriously as we take our weddings!—every misstep and gaffe get blown crazily out of proportion.

You have to call things off.

It doesn't happen only in movies. Here's a checklist of what to do when your beloved becomes behated.

1. Divvy up responsibilities with your former fiancé. This is going to be one serious to-do list (if you're still speaking, that is).

2. Phone your close family members and the people in your bridal party to let them know that the marriage has been called off.

3. Phone guests who will be attending the wedding from out of town, as they will need to

cancel flights and reservations. Then phone or write the rest of the guests to say the event has been canceled.

4. Make a list of all the services and locations that need to be notified. This includes, but is not limited to, photographers, musicians, florist, caterers, limousine services, the church, the reception hall, and a travel agent to cancel the honeymoon.

5. Contact all the services in your list to arrange to cancel the wedding and services. You may face penalties for the cancellation, and some services or items that you purchased may require that you pay substantial penalties. Although you might be stuck with your wedding gown, most other services will charge you a percentage of their fee for cancellation depending on the cancellation rights in your contract.

6. If you've purchased wedding insurance, now is the time to contact your provider (see sidebar).

7. Return all gifts (including your shower gifts).

8. Regardless of the circumstances, take the high road and return the engagement ring to your former fiancé. Why would you want such a sparkly reminder of what went wrong anyway?

Wedding Insurance

What it does: Protects you from the problem of extreme weather, no-show vendors, alcohol-related incidents, and calling it off.

Why you need it: It will cover your losses if bad weather, airport delays, or a sudden illness prevents you or your intended from getting to the church on time. If the local health department shuts down your caterer, most insurance policies will cover the extra cost of finding a last-minute vendor to feed your guests. And if the bridal store files for bankruptcy before you pick up your Vera Wang, wedding insurance will cover the cost of a new dress thanks to the "lost or damaged formalwear" clause.

Where to find it: WedSafe.com is an online company dedicated to helping brides and grooms cope with potential losses on their big day. Traditional insurance carriers such as Traveler's and Fireman's Fund also carry options for wedding insurance, and the latter provides "change of heart" coverage, protecting the innocent party in the case of a runaway bride or reluctant groom.

Your future mother-in-law wants to wear a sexy red dress to the wedding. And yes, she's still married.

Take the time to involve your mother-in-law in the wedding planning process so she knows all the details of your big day. Traditionally, the mother of the bride selects her outfit first, and then the mother of the groom chooses hers accordingly. Generally, the mothers should go for colors that blend in with the wedding party and never wear all black or all white. In order to nip a potential wedding faux pas in the bud, take matters into your own hands and offer to take your mother-in-law dress shopping—if nothing else, you'll at least be able to drop subtle hints about her wardrobe. When all else fails, beg your fiancé to change his mother's mind.

A friend or family member is hurt you that haven't asked her to be in your bridal party.

It is nearly impossible to avoid leaving someone out. Offer her the opportunity to be involved in the ceremony in a different way, as a reader or a greeter. If either one of these feels like a slap in the face, ask her to be an honorary bridesmaid, meaning that she will have special seating up front and her name in the wedding announcement.

You're stood up at the altar.

This is the stuff of season-finale cliffhangers for a reason. Horrifying. Criminal. Nightmare. Punishable by death (or at least castration). As difficult as it may be to maintain any sense of composure, you must choke back your tears, thank your guests for their time, and walk away. Once you're out of the spotlight, you'll be able to sort through your feelings with your close friends and family—and sew a life-sized voodoo doll out of his old shirts. Remember, the person who truly looks bad is the groom, not you. And while it is horrible to be left at the altar, it would have been worse to be married to the type of guy who would leave a girl at the altar. Plus there's no messy divorce.

Now, if *you* turn out to be the runaway bride, well, I'm not here to cast judgment. But I will say that showing up and explaining is far better than standing someone up entirely. And while you owe your fiancé and his family a verbal explanation (e-mail won't do here, my friend), simply tell your guests that you and the groom have decided to postpone. Let him save a little face, please. No need to go into any shortcomings publicly. You should also offer to reimburse your family or your former fiancé's family for any expenses they have paid, if you can. And next time make *sure* you're getting married for love, not for the ceremony or gifts.

You've gained or lost weight and your dress doesn't fit.

Don't worry—your tailor will offer a final fitting the week before you head down the altar and will probably be able to create a perfect fit. If you put on serious pounds just a couple of days before the ceremony, though, focus on debloating—cut out all white foods (carbs), drink tons of water with fresh lemon, and take long, long walks to get your digestive system working. Most important, invest in an amazing foundation garment (aka girdle). There's not a celebrity out there who sets foot on the red carpet without her Spanx nude bike shorts. If it's still a no-go, consider a ready-made backup from places like J. Crew. Okay, it's not your dream dress, but clearly, something had to give.

If, because of anxiety-induced weight loss, your dress has become loose in all the wrong places, recruit your bridesmaids to tailor with lots of safety pins and hem tape. Stuff your bra with silicone chicken cutlets and purchase a curve-enhancing padded girdle. (Yes, there is such a thing as a padded girdle.) And by all means, eat whatever you want.

You don't know whom to ask to be your bridesmaids.

The number of bridesmaids you have in your ceremony depends on its size and style, ranging from none at a small informal wedding to as many as fourteen for a large formal wedding. The honor of being your maid or matron of honor should be bestowed upon someone very close to you, like your sister or BFF. The makeup of the rest of your bridal party is up to you and should be a mix of the close friends and relatives (or soon-to-be relatives, like your fiancé's sister) who are most important to you. Don't feel pressured (like I did) to ask everyone who asked you to be in theirs.

One solution is to forgo bridesmaids and have lots of flower girls. Another is simply to say you know how expensive it is and didn't want everyone to have to fork out the cash. Or if you can, blame it on your mom and say she limited you to a certain number, and you drew straws.

You're worried about having your period on your wedding day.

If you're taking birth control pills, you can skip your period and have a stress-free white wedding. By beginning your next pack of pills during the start of your placebo week, you trick your body into not menstruating. Although there's some potential for spotting and/or breakthrough bleeding, skipping your period by the birth control method is perfectly healthy. Ask your doctor about Seasonale, the first FDA-approved extended-cycle birth control that allows you to menstruate just four times a year.

Remember that movie *27 Dresses*? Well, that gal had nothing on you. And you don't want to be a bridesmaid this time.

It happens to the best of us: you get a call from your summer camp bunkmate out of the blue and the next thing you know you're standing next to her in a royal blue dress. Oy. Unless this person is a very close friend or relative, do not feel compelled to do something you really do not want to do. Being a bridesmaid takes effort, money, and commitment, and if you question your ability to give any of those three things, do everyone a favor and politely decline. Simply tell the bride you're honored, but you don't feel you can commit to being a bridesmaid. As soon as you give her a reason, she'll start looking for a way around it, so let her know you're so happy for her and are obviously excited to attend the wedding as a guest, but you think it's best that she ask someone else to stand with her. As a compromise, you can offer to do a reading or take on some other task at the wedding to show the bride that you're supportive. And get her a fabulous gift—no matter what you choose, it will cost less than throwing her a shower.

You don't have the cash to attend a destination wedding.

Destination weddings are both a blessing and a curse. What could be bad about spending a weekend in an exotic locale with friends? Its effect on your bank account. By choosing expensive, out-of-town locales, a couple politely weeds out casual friends, so don't feel guilty for declining if you're not in the inner circle. If you're in the wedding party but unable to spring for the trip, politely ask the bride to suggest other guests with whom to share the financial burden—she'll know who would be willing to share a room or a rental car. Or stay at the hostel across town (which would probably be more fun than the five-star resort anyway).

You sleep with a groomsman.

Who can blame you? Love is in the air, champagne is flowing, and you're surrounded by a phalanx of eligible young men. If you get caught up in the moment and find yourself horizontal in the coatroom, be prepared to handle the aftermath like an adult. Even if you're into him, keep the affair under wraps until the weekend is over. If, the morning after the indiscretion, you find that your champagne goggles played evil tricks on you, avoid eye contact for the rest of the weekend and keep your distance. If he's persistent and making you feel uncomfortable, cut your losses and remove yourself from the situation entirely by feigning illness or family emergency. You can always tell the bride the truth later.

How to Stop a Wedding

When the first time you know you're truly in love with your best friend is the day he's set to marry someone else . . .

1. Make sure the odds of winning him are favorable. If he's ever uttered the words "I just don't think of you that way," grin and bear the nuptials.
2. Be discreet—wear all black and do not enter the ceremony site until just before the wedding begins.
3. Get in position at the back of the room.
4. When the clergy member or judge asks if anyone objects to the marriage, step forward, confidently raise your hand, and say, "I object," followed by a reason why your lover must not marry the other person.
5. Wait. Do so confidently and quietly. But you'd really better have a very good reason.

You're asked to give a toast—and you hate public speaking.

There are just two things to remember when giving a toast: don't drink too much beforehand, and know when to stop speaking. My own short-and-sweet toast recipe begins with a reference to my relationship to the bride and ends with a positive pronouncement about the couple's future good fortune. Funny anecdotes are always welcomed, but steer clear of embarrassing stories. And unless the wedding is taking place in Las Vegas, always keep it PG.

You think your friend is marrying the wrong guy.

Unless you have specific knowledge of extreme circumstances—violence, confirmed cheating, evidence of scandals—it's best to keep your mouth shut. You don't want to be the one to say "I told you so" when the marriage falls apart—or embarrassed to look her husband in the eye when they're still happily married forty years from now. Just make sure you have talked to her about marriage and give her the chance to back out if she jumped on a train she actually wants to get off.

6. If he objects, quietly exit the room and walk away without making a scene. Go into hiding for six months. Write a novel or travel the world.

7. If he turns away from his bride and comes loping down the aisle toward you, make a fast getaway by jumping on a city bus like Dustin Hoffman and Katharine Ross in *The Graduate* and get ready to endure the wrath of a lot of people.

PART 3

Sticky Situations at Home

Your mom had your dad; Murphy Brown had Eldin. But all you've got is a box of pink tools. Once you try to unclog your own drain, you realize there's a reason plumbers get paid $300 per hour.

More and more women are nesting on their own these days. Based on my entirely nonscientific research, this has resulted in an epidemic of squeaky doors, leaky faucets, drafty windows, mildewed shower ceilings, dead smoke detectors, slow-draining bathtubs . . . the list goes on and on.

Wake up, people! If you got a big spaghetti stain on your vintage silk Halston, would you leave it to rot in the closet? I didn't think so. So, you shouldn't let the closet rot, either. It's probably worth more than the dress.

In the Kitchen

Confession: I was twenty-five before I stopped storing books in my oven. And clearly I'm not the only one. I have a friend who was once cooking Thanksgiving dinner for her entire family in her New York apartment. Since she didn't have enough oven space, the people next door, who were away, were gracious enough to lend her theirs. She ran across the hall to preheat the oven, then came back to her own kitchen to do her prep work. When she returned to her neighbors' apartment, it was on fire! Turns out I wasn't the only one storing books in my oven. And while she was able to salvage both the apartment, the friendship, and even the meal, it has made her a bit oven-shy about both borrowing from others and cooking without being extra careful.

What is it about being a Bottom Chef that makes us feel so bad? It probably has something to do with the way Western society has defined femininity since the beginning of time—any woman who can't cook isn't really a woman.

Luckily, there are some easy cheats that will fool everyone—and maybe even help you develop a fondness for that shiny room full of strange appliances.

You dropped your roommate's supercool cocktail ring down the garbage disposal—and you may have forgotten to mention you borrowed it in the first place.

First unplug the disposal. This will prevent the important item—and your fingers—from getting digested. Then go to a flower shop and get some floral clay. You can use some Play-Doh if you have it around, but the advantage of floral clay is that it doesn't fall apart when wet. Stick a lump of it on the bottom of a wooden spoon handle. The ring (or bottle cap, or whatever) will stick to it, and you won't have to waste $200 on a repairman because you let your InSinkErator chew up a piece of metal.

You have six guests coming over for pasta, and your gas burners won't ignite.

Light a match, hold it over the burner, and slowly turn on the gas until it ignites. (Tie your hair back, and don't get too close!) If there's no gas, you need to call your gas or appliance company.

Your perfectly composed dish of eggplant parmigiana now looks like it exploded in the oven. You don't have three hours to put your oven through a self-cleaning cycle.

You don't need to use a toxic oven cleaner. Just let the oven cool, wet the spill, and dump a bunch of ordinary table salt on top of the messy spots. Let it work for a few minutes, then scrape the salt away and wash the area clean with a warm, wet rag.

You left a wad of plastic wrap on top of the toaster—and now it's melted on.

Unplug the appliance and let it cool. Dip a cloth in nail polish remover and wipe over the spot, then wipe off with a damp cloth.

Your fridge is leaking—and you barely keep a thing in it. So what's leaking?

Leaky fridges typically happen only when you have an icemaker. Tighten the fittings in the back of the refrigerator where the water goes into the icemaker. If it's not a self-defrosting fridge, you need to take everything out, unplug the fridge, rinse out the inside, and use a hair dryer to melt the ice in the freezer. Sorry, I know this doesn't sound like fun. Don't blame the messenger!

You can't get the blender clean, and your nosiest, cleanest friend is breathing down your neck as you make the margaritas.

Make a sudsy shake. Fill the pitcher with warm water and a few drops of dish soap and buzz away for ten seconds. Wait ten seconds, then buzz again. Repeat until all the grime is loose.

If the blender still seems a bit scuzzy, pass it to your friend and say, "How lucky for me that you're here! I know you must know how to get this thing sparkling—would you mind?"

Your drain is clogged, but you don't want to pour toxic sludge down your sink.

Pretend you're back in fifth-grade science class—pour 1 cup of baking soda down the drain, then chase it with 1 cup of vinegar. Then block or cover the drain so that the force of the gas created goes down the drain toward the clog rather than up and out, where it won't do any good. Let it sit covered for five minutes, and then follow with a gallon of boiling water. Another way to unclog your drain chemical-free is to feed it ½ cup of salt, followed by boiling water. Continue to flush with very hot water until the clog breaks.

Your mother-in-law is coming over for dinner—and her family's china is stained beyond belief.

When dishwashing liquid won't remove years and years of stains from good china, soak it in Gatorade (the active ingredient is citric acid).

You have nowhere to store ice for a party.

Put it in your washing machine, along with anything else that needs cooling. Just don't forget to remove the beers before running the lingerie cycle.

You can't open a jar (and you tried the hot-water thing).

Punch a hole in the lid with an ice pick or strong knife. This will break the vacuum seal. If that fails, a tough jar is the perfect excuse to pay a visit to that Adonis next door.

You're opening a bottle of wine—the most expensive bottle of wine you've ever held in your hand, brought graciously by your guests—and the cork breaks.

Instead of pushing the remaining cork into the bottle, carefully work the worm (the spiral part of the corkscrew) back into it. If the cork crumbles into the bottle, carefully strain the wine through cheesecloth or a coffee filter. Strain the whole bottle into a decanter and your guests will have no idea what's happened.

You made fried everything, and now your walls are covered in grease stains.

Make a solution of 2 cups water and 2 tablespoons dishwashing liquid, dampen a rag with it, and use the rag to rub stains away. This also removes greasy fingerprints from door frames and other painted woodwork.

You've been cooking with onion and garlic and now your hands smell like an Italian restaurant.

Rub your hands against anything stainless steel that's wet. A sink or a spoon will do the trick. Then wash with lukewarm water and dishwashing soap.

It's suspected that the sulfur molecules that create the distinctive onion smell react with the metals in the stainless steel and become neutralized. There are also special kitchen soaps made for this purpose—look for those containing citrus or tea tree oil.

You've been cooking with onion and garlic and now your house smells like an Italian restaurant.

Well, that's not always a bad thing! But if the smell lasts long after the dishes are done, just simmer some white vinegar in an uncovered pot on your stove for half an hour, then remove. To freshen the air in your kitchen, especially after cooking fish or cabbage, place a whole, unpeeled lemon in a 300°F oven for about fifteen minutes, leaving the door slightly open. Turn off the oven and let the lemon cool before removing it.

Something spoiled in your fridge . . . years ago. But you're still haunted by the smell of its ghost.

Sometimes baking soda isn't enough. Here's a heavier-duty solution: Keep charcoal briquettes in the fridge to sweeten the air (but avoid briquettes impregnated with lighter fluid). You can also saturate a cotton ball in lemon extract and place it in an open container on the bottom shelf.

Your microwave smells like hot dogs . . . and you didn't even make hot dogs!

A mixture of 1 cup water, the juice and peel of 1 lemon, and a few cloves placed in a glass bowl will neutralize bad odors in your microwave. Just place it in the microwave for three minutes on high, then let cool and wipe the walls of the microwave with a paper towel.

First Aid: When the Emergency Room Is Too Much, but a Band-Aid Just Won't Cut It

You sliced your thumb.

Since human saliva speeds up wound healing, follow your instincts and stick your finger in your mouth.

The more than two hundred compounds found in saliva have proved capable of preventing cavities, fighting fungi, and even protecting against HIV transmission. Most kitchen nicks need no more than a few minutes of pressure to stop the bleeding, and then the application of a Band-Aid or, even better, a liquid bandage (I like New-Skin, which you can get at Target). Head to the hospital for stitches if the cut is longer than an inch, won't stop bleeding, or seems very deep.

You burned your arm.

If your skin is red, sore, and a little bit swollen, you have a first-degree burn. Only the top layer of skin is affected, and the skin doesn't blister. To treat a first-degree burn, soak skin in cool water, then apply aloe vera gel or, if the skin peels, an antibiotic ointment such as Neosporin. You can cover the burn with a loosely wrapped gauze bandage and take ibuprofen for mild pain. A second-degree burn reaches deeper than a first-degree burn. It causes blisters that are filled with clear fluid. Treat second-degree burns the same as first-degree ones, except be sure to change their bandages every day, since blisters ooze. (Sorry if that's gross, but it's normal.) If you notice any signs of infection, such as pus, increased swelling, or extreme, localized pain, head to the doctor immediately. And note: An old home remedy many of us have heard about—putting butter on a burn—can do more harm than good, because the butter, which isn't sterile, increases the risk of the burn becoming infected.

You swallowed Drano or something similarly toxic.

Poisons are classified into two different categories. Corrosive poisons, such as hydrochloric acid or sodium hydroxide, damage flesh. If you swallow one, do not vomit, as the poison will be as damaging coming up as it was going down. Instead, swallow milk, an egg white, or a mixture of flour and water, and call 911. Noncorrosive poisons don't damage flesh but have harmful effects on the nervous or circulatory systems. (Snake venom and arsenic are noncorrosive.) If you swallow one, try to vomit, then drink 1 or 2 tablespoons of activated charcoal dissolved in a glass of water (might as well have some on hand) and head to the hospital.

The Modern Girl's List of Kitchen Catastrophe Must-Haves

Floral clay—to retrieve lost items from the disposal

Cheesecloth—to filter impurities out of most any liquid

Charcoal briquettes—to eliminate odors

Antibiotic ointment—for burns and cuts

Denture-cleaning tablets—to remove red-wine residue from decanters and glasses

Baking soda—to extinguish small grease fires

Lemons—to deodorize a stinky garbage disposal

Chocolate—to make you feel better after dealing with whatever disaster has befallen you

While checking on the broiled chicken, you scorched your eyebrows and eyelashes.

Fake the appearance of eyebrows with a stiff eyebrow brush dipped in brow powder (you can also use an eye shadow a couple of shades lighter than your brows) and feather it up in the direction of hair growth. Be sure to use a light touch. False eyelashes look great and are easy to apply. You may need to trim them to be sure they fit your lid—too-long lash strips irritate eyes and look fake. Be sure to use a hypoallergenic adhesive if you're going to be wearing them every day. Lashes take a couple of months to grow back, but if you're eager to have results sooner, try a product called RevitaLash. It was originally invented to help glaucoma patients, but doctors discovered it helps eyelashes grow faster and longer, and now it's being sold as a beauty product.

In the Living Room

You spilled red wine on your white couch.

It was bound to happen, and it's not the end of the world (or the upholstery). Blot the stain with paper towels immediately. Don't rub. When all the moisture has been absorbed, use a sponge to saturate the area with warm water. Then apply a heavy coat of talcum powder and allow the powder to absorb the water for a few minutes. Use a brush to remove the powder, and the stain should go with it.

If this method doesn't work, you can rub salt into the spill as though you're exfoliating the couch, then leave it on overnight. In the morning, vacuum up the salt and repeat if necessary.

If your couch is slipcovered, pretreat the stain with a mixture of blue dishwashing soap and hydrogen peroxide. (Remember, it's just for whites, though.) Add just enough peroxide to the dish soap until it turns clear, then apply to the stain, which, according to devotees of the formula, will disappear before your eyes. Then launder as usual.

In a heated midnight furniture reorg, you dragged the sofa across the floor and left a trail of scratches.

Remember those almonds and walnuts you bought when you committed to eating healthy snacks? Break a few in half and rub the broken end of the nut over the scratches in a circular motion. The nuts' meat and oil will camouflage the scratch and fill it in. Seal with a coat of old-fashioned floor wax.

Spending the weekend at your boyfriend's parents' house, you leave a huge water ring on the antique coffee table.

Wow, way to make a good first impression. Don't panic. Squeeze white toothpaste onto a damp cloth; rub along the grain of the wood, then polish with a clean cloth. If you spoil your pearly whites with fancy bleaching toothpaste, buy a cheap tube of Colgate just for this purpose—the same ingredients that whiten your teeth could whiten the table and make things worse.

You spilled candle wax on your floor.

Take a large brown envelope or a paper grocery bag and a hot iron. Place the paper over the wax stain, then press the hot iron on top of the envelope. The iron heats up the candle wax and the brown paper absorbs the hot wax. A trick that works on both wood and carpet! In the future, if you want drip-free, long-burning candles, put them in the freezer overnight before burning them. And always choose soy candles over wax if you have a choice—they offer a cleaner burn than petroleum-based waxes and don't contain carcinogens.

Your ungainliest male friend threw himself onto your sofa and broke the springs. How, exactly, does one replace sofa springs?

I'm sorry to have to be the one to tell you this, but it's not worth it. Replacing sofa springs requires taking apart the furniture entirely—removing the upholstery and the batting—and then putting everything back together again. You certainly couldn't attempt such a thing yourself, and the cost of hiring an expert to do it would be prohibitive. Head to Crate and Barrel, and make Shrek at least buy you a nice dinner.

You caught your stiletto in the edge of your great-grandmother's Persian rug and ripped off a long piece of yarn.

Provided this is not a museum-quality piece—did you know there are rugs that cost a million dollars?—you can sew the thing back together yourself. Just get some nylon thread and a curved needle and stitch the separated piece back onto the body of the rug. Make the stitches as small and close together as possible and no one will even notice.

You dropped a glass and it nicked the edge of your glass coffee table. Sure, the chip is small—but it drives you crazy every time you walk by it.

Just paint some supershiny clear nail topcoat on the chip. Keep adding layers until the surface feels smooth. If you don't like the way it looks once it dries, you can always remove it with nail polish remover.

Your new puppy thinks that since your couch is green, he should pee on it.

Having been the proud mama of many animals—right now Marshmallow, aka Clifford the Big White Dog, is my nonhuman baby—I am quite familiar with the smell of puppy pee-pee. No matter how many times you scrub a chair or couch cushion, the stench always seems to remain. That's because the odor has bled down beneath the surface of the fabric into the cushion. In order to truly remove it, you'll need to unzip the cushion from its cover and soak the foam in odor remover—preferably an all-natural one. Look for solutions that are non-toxic, fragrance-free, dye-free, phosphate-free, and also biodegradable. If only I had bought stock in Nature's Miracle when I first got my dog, I'd be a rich woman by now.

You knocked your lamp over and it's not working anymore.

Okay, you shook the bulb and replaced it if it rattled, right? And you made sure the plug didn't get knocked out of the outlet? How about being sure the outlet's active?

If your lamp still isn't working, check to be sure the switch goes back and forth smoothly and that the cord hasn't separated from the base. If these things seem to be in order, head to your local hardware store—it's best to have your lamp in tow, if possible—and purchase new wiring. The salesperson will be able to help you, and the whole thing shouldn't cost more than about $10.

In the Bathroom

You spent the night at your boyfriend's house and somehow managed to clog his toilet.

The good news is, he left for work. The bad news is, you've got to deal with the clog. Assuming you didn't shove photos of his ex-girlfriend down there, in most cases a regular plunger will do the trick and get everything flowing. However, sometimes old plumbing and/or a huge clog makes it difficult, and the next step is to go to a hardware store and buy an auger—a curved metal object that can be used to clear pipes of blockage.

If your toilet is clogged due to a solid object, squirt a generous amount of dishwashing liquid into the clogged toilet bowl. Wait until the water level goes down a bit. Then fill a bucket (or empty wastebasket) with water and pour in from about waist high. You may need to repeat this two or three times, but odds are that even the first or second time you'll hear something starting to happen. No mess, no broken seals, no overflow (assuming you stop pouring from the bucket before the water goes over the top).

You've dripped brown hair dye all over your white bathroom.

Whatever you do, don't let it dry. Rubbing alcohol will remove dye from marble and tile. A solution of vinegar and water will remove it from fabric, such as your shower curtain or your bra strap. Throw your clothing and towels into the washing machine with detergent, ammonia, and warm water.

You spilled red nail polish all over your white bathroom.

In this case, let it dry. Then just gently flick it off with your nails. This method works on shoes and handbags, too—so it's safe to say you shouldn't cry over spilled polish. Just ask my daughter, who cowered in fear when she and her friend dumped an entire bottle of Chanel red all over my new marble, and then lived to tell the tale when I was able to clean it up. You can also try using nail polish remover on hard surfaces—though not fabrics.

Your grout is brown and disgusting, no matter how hard you scrub it. You don't want guests to think you're dirty!

Instead of scrubbing with a toothbrush and toxic bathroom cleanser, just spray some foamy shaving cream all over the tiles and grout and massage

Real Genius? Weird Science?

These household products will—sometimes bizarrely—save the day.

- An envelope moistener removes the dust and dirt from the leaves of houseplants.
- A soft-bristled paintbrush is the perfect dusting tool for a pleated lampshade, allowing you to clean between tight ridges and grooves.
- A piece of chalk, used to draw a line along windowsills and baseboards, will keep ants from entering your house.
- WD–40 will instantly remove adhesive from skin.
- Baby powder silences squeaky floors and makes stubborn knots virtually untie themselves.

with a sponge, then rinse with your showerhead. Hydrogen peroxide works, too!

The sound of your leaky faucet is driving you insane.

There's a reason why they call it "water torture." This quick fix is not meant as a long-term solution because it doesn't address the environmental implications of a leaky faucet (which can waste up to twenty gallons of water per day), but it's a lifesaver when the *plip-plop* is keeping you up all night. Tie a two-foot-long piece of string to the faucet's nozzle, then put the rest of the string down the drain The string will carry the water down the sink silently.

You spilled hair removal wax on your bath rug.

Use a putty knife to scrape up as much wax as you can. Place a piece of wax paper and then a bath towel over the remaining wax. Iron over the towel with a hot iron, then lift up the towel and wax paper. The hair-removal wax will abandon the rug for the wax paper.

- Plain white toothpaste (nothing fancy—basic Colgate is best) polishes tarnished silver, brass, and bronze. It's also a soothing ointment for kitchen burns.
- Tennis balls will remove scuff marks from floors.
- Running a packet of original orange Tang drink mix—a favorite of astronauts back in the day—through your dishwasher's full cycle will remove soap scum and food buildup and bring it back to optimum function due to the citric acid and who knows what else.
- Petroleum jelly, applied to the inside of candle-holders, will keep wax buildup from sticking.

In the Rest of the House

You don't have A/C and need to cool down, ASAP!

In the summer, draw the shades or curtains to keep out unwanted sun. Open your windows for optimal cross-ventilation: the window on one wall should be lower than the one on the opposite wall, and the higher one should be opened widest to allow for faster airflow.

Try eating Ayurvedic cooling foods such as cucumbers, watermelon, and bitter greens. Avoid egg yolks, nuts, hot spices, honey, and hot drinks.

A window fan can work wonders on a warm summer night. Since outside air will almost certainly be cooler than the stuffy, humid air inside, the airstream emanating from the fan will feel like air-conditioning.

Sticky Note: Never turn a ceiling fan on reverse when you want to cool things off—because it draws cool air upward, it'll undermine your efforts entirely.

You have drafty windows.

Drafts from under doors or windows can be blocked by laying a handmade "snake" across the crack. Make a tube out of material or use a knee sock or several socks sewn together. Stuff the sock with any bulky material (dried beans or lentils work perfectly), and then sew up the open end.

A broken pipe is about to flood your basement.

Cut a patch of old inner tube to cover the leaky part. Secure with small hose clamps to keep things dry until a plumber arrives. What's an automotive clamp, you ask? It looks like a small vise. Buy some now and be ready. Might as well get some squares of rubber while you're at it.

You scratched the floors in your rental.

If a scratch cuts through the finish and into the wood, the piece will probably need restaining. Although you can use commercially available colored wax sticks, you will get better results with oil stain or acrylic artist's paint. First, clean the area with benzine. Then touch up the scratch to blend it with the surrounding area. Let dry for twenty-four hours and repeat if necessary.

Wood markers (and nuts, as mentioned) are good for spot treatment of shallow scratches. Another hardware store find is Howard Restor-A-Finish, an easy-to-apply product that helps wipe away scratches, water rings, and fading on hardwood floors.

Your floors are squeaky and you're beginning to feel like you're living in a B horror movie.

Squeaky floorboards can be silenced with a number of lubricants—powders, such as talcum or graphite, or liquid furniture wax. Just squirt a bit of the lubricant onto the joint through a nozzle.

You ruined the walls in your rental and you don't have time to jet to the hardware store before your inspection.

Toothpaste (white paste, not aqua gel!) is a quick and easy alternative to spackle. Just daub it into the holes and scrape off the excess with a piece of cardboard or a credit card.

You're locked out of your apartment and are in no mood to pay for a locksmith to let you in.

First, if you're renting but your name isn't on the lease, it may be illegal for you to break in. This is the kind of thing it's good to know ahead of time—so check the lease!

If you're in the clear, open any window that you know is not securely locked and crawl in. Don't break a window if you don't have the money for a locksmith—window glass is more expensive to replace (about $500 on average) than a lock.

If you don't have an open window, you may be able to use a flexible piece of plastic (like a credit card) to push back the lock on your door and break in. If this works easily, you should look into installing new locks—if you can do it, so can everyone else.

In the future, keep a spare key in your car glove compartment and one in a secret hiding spot in your yard so you are never left in such a bind. As a general rule, I like to make three extra sets of keys when I move into a new place: one goes into my glove compartment, one I give to my best friend, and one is hidden somewhere outside the house. Few things are more frustrating than having to fork over hundreds of dollars for the privilege of entering a place you already pay for.

When Your Home Is No Longer Your Own: A Crash Course on Vermin

You're overrun by ants.

Ants, like teenage boys, are always in search of junk food. To prevent an ant-festation, store foods in tight containers, keep pet food bowls off the floor, and immediately wipe up spills and crumbs with equal parts vinegar and water. If you're already infested, you have a few options.

Go straight to the source. Try to locate the ants' nest and see if they can be cut off at the pass through repairs, by caulking, or by eliminating sources of moisture. Note the surfaces they crawl over, where the ants come from, and where they go after feeding—this will help you determine if they're nesting inside or coming in from outdoors. Sprinkle baby powder at possible entry points—windowsills, door frames, holes in baseboards. Or, as I mentioned earlier, you can use regular white chalk to draw a line to block their way. Ants, unlike professional skiers, detest white powder.

If you can't figure out where the ants are coming from, you can also use boric acid traps to seduce and then poison them. Be careful, though—boric acid is hazardous to children and pets.

You're bugged by bedbugs.

Contrary to popular belief, having bedbugs doesn't mean you're a filthy little piggie. The miniature vampires punish you anyway, though, sucking your blood while you sleep. Then they pretty much ruin your life. Like teenage fashion models, they can live up to a year without feeding. The oily liquid they produce is stinky and pervasive, and they leave tiny spots of bloody poop on your sheets.

Now, let's take a station break for a second. This. Is. So. Gross. I am so sorry if you're reading this section because you actually have bedbugs, and not just out of morbid curiosity. I would invite you to stay at my house if I could, just to get you away from them. Actually, maybe I wouldn't, because then my house might become infested. I'd get you a hotel room. At a hotel I never, ever want to stay in.

To avoid bedbugs, never, ever take upholstered furniture off the street—especially if you live in New York City. Buy secondhand pieces only from reputable dealers.

To banish these disgusting, bloodsucking miscreants, you must first find their dirty flophouses. Look for bloody spots and teensy eggshells on seams of mattresses and tufts of pillows. You may also find them behind loosened wallpaper or baseboards, in plaster cracks, and nestled in between pages of books. Remember that novel you used to read once a year? Yeah, not anymore.

Vacuum every area of your house you think

they may have set foot in. Repair and fill cracks and crevices. Run every single item of clothing and linen through the dryer at the highest setting to kill the bugs. Once you vacuum your mattress, cover it with a zippered cover and tape over zippers and openings to trap leftover bugs inside. Bedbugs can live a long time without a meal, so you should leave the cover on the mattress for at least a year. Or buy a new mattress. Bummer.

The fleas won't flee.

Fleas are the man whores of the insect world—they jump from host to host and transmit diseases (typhoid, tapeworm—even the plague, unlikely as it is). Fleas catch a ride into your house via your pets, but they'll only stay the night if your house is dusty, so it's time to pull out that vacuum. Groom your pets regularly and wash all their stuff with flea-killing soap.

You want to block the cockroaches.

You think Octomom is an impressive spawner? A single female cockroach can bring thirty thousand little roachies into the universe over the course of just one year. Cockroaches love all the same foods we do, as well as book bindings, fabric sizing, shoe soles, and dead insects. They pick up germs and leave them behind wherever they walk. They also poop out toxins they've digested, and their poop stinks, even after they're long gone.

Keep a clean, dry house and cockroaches will stay away. You can also place tight screening over any small holes in your home to keep them from crawling through.

Bay leaves drive away roaches. Place them in the pantry, in cupboards, and on shelves. A compound found in cucumber is also an effective cockroach chaser—place their skins wherever roaches roam.

There are moths everywhere, and it's creeping you out. Plus, there's the matter of cashmere sweaters . . .

Moths live in dark, undisturbed areas such as closets, basements, and attics, and tend to congregate in the corners or in folds of fabrics. They produce eggs that hatch into larvae that feed on animal-based materials, including wool, hair, fur, silk, felt, and feathers. Synthetic materials are rarely attacked unless blended with wool or unless they're dirty. So before packing these items for storage, wash or dry-clean them.

Moths do not like the smell of cedar. Or suspend sachets filled with lavender (and/or laced with its oil) from closet rods or tuck them into your drawers. Lavender will not kill existing moth eggs or larvae—it's simply a good way to prevent an infestation. Stay away from mothballs and moth crystals; they contain pesticides that can be harmful to people and pets.

Dust mites might drive you batty. (At least you don't have bats.)

Dust mites feed on skin flakes, hence they love to nest in warm, moist places like the insides of pillows and mattresses. While they're not dangerous, dust mites are a common cause of asthma and allergies. A typical mattress may contain up to two million dust mites! That's as many people as live in Las Vegas.

To minimize reactions to dust mites, get dust mite covers for your pillows and mattresses. Opt out of carpet—its fibers can trap skin cells and attract mites. Wash sheets at least once a week in hot water, and use a dehumidifier to keep bedrooms dry.

Your pad is a fly trap.

Since I doubt (and hope) you don't have piles of manure, rotting flesh, and garbage around your house, you can check off your list the number one way to minimize flies. Also, don't leave food out. While flies don't have teeth, they still manage to eat ridiculous amounts of food by spitting on it to predigest it, then sucking it in. A single housefly can carry more than thirty million bacteria internally and another half billion outside its body.

Flies hate vinegar and eucalyptus. Maybe that's why old people seem to avoid being plagued by them. Just make your house smell like a nursing home and flies won't come near you.

You're sick of living on the mosquito coast.

Unlike the rest of us, mosquitoes love heat and humidity. Keep it cool and dry and they won't bug you.

DEET is typically the most effective mosquito repellent, but it's been known to cause side effects such as rashes, dizziness, and confusion when used in high concentrations. Many people prefer to use citronella candles, but their efficacy is debatable. Mosquitoes will stay far away from your summer barbeque if you throw a piece of garden sod or a handful of green grass into burning charcoal. Just don't do it while the meat is still on the fire, unless you're looking to create a bizarre flavor fusion worthy of *Top Chef*.

You don't need to go fishing for silverfish—they're everywhere.

Silverfish love dark, warm, moist environments such as attics, closets, baseboards, and around bathroom fixtures. Like me, they're obsessed with carbs and enjoy gorging on starch from wallpaper, books, and fabric (I prefer macaroni and cheese).

They leave little holes in paper and little yellow spots on fabric. Ant and roach sprays will kill silverfish, as will boric acid.

You're being kissed by the spider man, woman, and child.

Most of us mistake all big, leggy bugs for spiders, but a spider's not a spider unless it has two body segments, eight legs, and no chewing mouth parts (at least you know a spider won't chew you). Contrary to popular belief, most spiders are not big or angry enough to bite humans. They're total hermits, too, so they hate you as much as you hate them.

The best way to manage spiders is by vacuuming up their webs. Pay special attention to windows, ceilings, and corners. Slivers of eucalyptus-scented soap deter spiders, as does pennyroyal oil or rubbing alcohol on household surfaces. You can also spray webby areas with a solution of 8 ounces of water and 3 tablespoons of liquid soap.

Being in your kitchen at night is like watching *The Nutcracker*—the mice are large and in charge.

Mice enjoy spending the holiday season indoors—they tend to invade our homes between October and February. You'll know they're paying you an extended visit if you see little teeny footprints everywhere. Or droppings. Mouse traps do snag mice when baited correctly, but don't use cheese—try peanut butter, raisins, or whole-grain bread instead (mice are health food fanatics). And don't touch the traps with your bare hands; wear gloves. Mice won't approach the trap if it's redolent with the essence of their landlords. You'll need more mousetraps than you think—five to ten for every mouse hole. Place them at right angles to walls.

Now that we've solved all panic-inducing—okay, sometimes even life-threatening!—issues, it's time to tackle some less pressing but equally important predicaments.

Feathering Your Nest

You need to furnish your new place, but you're broke. And cheap furniture is so depressing!

Those of us with more taste than money can pick up high design at low prices from stores like Ikea, Target, West Elm, and CB2. Just examine the floor models for quality (for instance, you might want to choose solid pine bookcases over cheap but chip-prone veneered particleboard) and recruit a friend to help you with the assembly.

For colorful, mod designs, try shopping outside your demographic at PB Teen, Delia's, or Urban Outfitters. Skip too-juvenile patterned fabrics (do you really want a Hawaii-themed bedroom?) in favor of clean, graphic linens. Hint: the boys' section often has the most understated offerings.

Prices for antiques have recently plunged: check out Craigslist and your local newspaper to find estate sales. And don't turn up your nose at thrift shops—alongside the velour couches and flimsy dinettes are solidly built pieces that are just a coat of paint away from chic. Avoid buying upholstered items unless they come from a very reliable dealer—no bargain is worth bedbugs.

Comb eBay for lighting fixtures and unique accessories. Factor the price of delivery into any bargain—consider shopping through a single dealer for combined shipping or look for sellers in your area.

Look for leftovers: granite and marble suppliers often have remnants that could be used to update a bathroom floor. Tiles left over from bigger jobs can also be bought at a discount—consider mapping out a whimsical mosaic with multihued tiles, (carefully cut) pieces of mirror, and ceramic curios. You can also hit the hardware store for cut-rate paint returns and spare rolls of wallpaper; the key is to be open to the colors you find.

You have a teeny-tiny living space and want to maximize it.

The key is to rethink the way your space is laid out. Turn your walk-in closet into a secret work zone by installing a makeshift desk. Or if the closet is (just) big enough for your bed, turn it into a cozy cocoon and double the size of your living space.

Choose double-duty furniture—like bed frames with drawers built into the base—since space is at a premium and storage usually insufficient.

Reconsider the coffee table. A pair of round, glass-topped steel end tables can take the place of a traditional rectangular monolith.

Furniture that folds down, pulls out, or moves around can ease the problems of limited living quarters. Murphy bed, anyone?

Why Not DIY?

- Use a few coats of matte white spray paint to reinvigorate an assortment of 25¢ vases and other flea market finds. Group the finished products together or mount them individually on sconces for a look that's very Jonathan Adler–esque.
- Make your own "curtains" by threading a curtain rod through the hem of a flat sheet (just cut a slit on each side to form a tunnel). Patterns you'd never consider putting on your bed can look great as window coverings, and it's fine to buy inexpensive, poly-blend fabrics since these sheets will never touch your skin.
- I fell in love with a friend's spectacular feather-covered lampshade and was shocked when she told me she'd made it herself. Using a hot glue gun, she layered hundreds of feathers from a craft store over an inexpensive fabric lampshade from Urban Outfitters and used it to top a lamp base she

found for $1 at a thrift store.

- If you have an old dress or coat in a dated silhouette but amazing fabric, make pillows! Just cut two big squares, sew them together with a machine or by hand (if you do it by hand, be sure you make tediously tiny stitches so the stuffing doesn't creep out), and fill with fabric batting (you can get it for next to nothing at a craft store). You can even combine two different fabrics for an eclectic look.
- I love the handmade knit throws you find at eclectic home stores such as Anthropologie, but they can run into the hundreds of dollars. You can make your own using just a pile of old wool sweaters, a washer/dryer, and a sewing machine. Just felt your sweaters by washing and drying them on high heat—this condenses the fibers and prevents sweaters from fraying—then cut them into rectangles. Sew them together in a random patchwork and you've

For a small kitchen, think about buying a table with a pedestal table base—it makes room for more legs under the table.

Max out vertical space by investing in high-style floating shelves that can house books and supplies.

Open up a small space by creating continuity between walls and floors, using the same color paint on the floors you do on the walls.

If you have large windows and if privacy isn't an issue, flood the space with natural light and avoid fussy curtains or shades.

Don't be afraid to use space enlargers to add dimensions to small spaces—this includes mirrors, wallpaper, murals, pale colors, and area-dividing screens.

You have a gigantic, cavernous living space that feels more like an art gallery than a home.

Hello? Don't you salivate every time you see an eighties movie starring a loft that outshines any of its characters? What girl hasn't fantasized about shimmying around Jennifer Beals' loft in *Flashdance*, or jumping up and down on Tom Hanks' trampoline in *Big*? Still, without access to Hollywood set decorators, it's easy to lose yourself in a never-ending space.

The first thing to do is paint. Choose rich, deep, highly saturated colors with a matte finish—deep eggplant, chocolate brown, and indigo blue are gorgeous—to visually shrink and cozify the space. You can layer on wall decals to add interest and distinguish open living spaces from one another. Wall decals are inexpensive, and you can always move them should they start to bore you.

Next, move the walls. Not literally, of course, but don't be afraid to create invisible walls with furniture, curtains and lighting. Place sofas in the middle of the room rather than along the edges of the room. Hang floor-grazing curtains from the ceiling to soften and divide space.

Then be sure you've got visual action happening on multiple levels. Mix luxe, puffy seating with low coffee tables and floor cushions. Try a patterned rug, and hang art at eye level. For high drama, be sure window coverings start at the ceiling and break at the floor.

Low, warm lighting is your friend. I'm talking table lamps and votives all over the place.

The best way, though, to make a massive place feel warm and cozy is to invite your friends over. Especially to show off all the improvements you've just made to your pad!

You have bad lighting. A girl has to look gorgeous in her own house—but how?

Lighting is something that can make or break the mood and style of a room, so replacing the weird, harsh ceiling fixtures that often come with rentals will do wonders for both the attractiveness of

got a cozy, designerish throw.

- Fashion a one-of-a-kind picture frame using the cover of a favorite childhood book and a hot glue gun. Just use an X-Acto knife to cut the front cover off the book, then cut a window in the cover for your photo to peek through. Tape the photo from behind to secure it to the window (cover it first with a plastic sheet protector if you're worried about its surface being exposed), then use the hot glue gun to glue three of the book cover's edges to a piece of wood the same size. Display it flat on your coffee table like an art book.
- Craft the world's cutest bulletin boards using vintage game boards. These can usually be had for a couple of bucks at yard and garage sales, especially if the game is missing pieces. (Don't worry, you won't need them anyway.) Just attach an old Scrabble or Monopoly board to a piece of corkboard cut to fit (Lowe's

or Home Depot will do it for you) using spray mount, then hang the board in your kitchen, entryway, or bedroom and collage on memories or reminders.

your room and the people in it. There is no lighting scheme duller or less flattering than the glare of an overhead light source (okay, maybe a fluorescent one, but hopefully you're not living in your cubicle). Here's a little secret every decorator knows: well-placed lighting can eliminate the need for new furniture or paint.

The Golden Rules to Achieving a Flattering Glow

There are three types of lighting. Background lighting washes the room with soft, general illumination. Task lighting focuses light directly onto a work surface. Accent lighting spotlights an important accessory or furniture grouping. The most finished-feeling rooms combine all three types of illumination.

Abundant natural light is always best, but even if you aren't lucky enough to live in a sunny space, you can fool yourself and your guests with a few simple tricks.

Combine natural light with artificial light so that each complements the other. To make the most of colors after dark, artificial light should come from the same general direction as natural light.

Dimmers are cheap and easy to install, and they can be used to create a variety of different ambiences.

Even in small rooms, aim for at least three light sources—ideally a floor lamp with matching table lamps—to banish gloomy corners.

As a general rule, table lamps cast more flattering light than overheads.

When lighting a room, think about variety—instead of one 200-watt bulb glaring harshly overhead, place three to four lamps around the room that each hold a 25-, 40-, or 60-watt bulb.

Lightbulb Moments:
Deciphering Wattage and Bulb Type

Wattage simply represents how much power is required to run a bulb. Lower-wattage bulbs require less electricity—and give off less light.

Standard incandescent bulbs give off a warm glow—and waste more energy than more earth-friendly options like LEDs or CFLs. This otherwise inexpensive option is best used for flattering mood lighting.

LED bulbs use less energy and give off a whiter light than incandescent bulbs. The bulbs are relatively more expensive but also last longer. Because the light can seem colder, try using the bulbs for reading or other task lamps.

CFLs or fluorescents are great for the environment—and while they used to give off a soul-crushing greenish light, they now come in warmer shades and can be used almost anywhere.

More on Decor

Your cut flowers tend to die in two seconds.

Choose flowers with buds that are tight, not totally open.

Add either a little salt, vinegar, sugar, baking soda, aspirin, or Sprite to the water.

When they start to wilt, spray the petals with a little hairspray. They'll freshen up just like a good bouffant.

For tulips, prick the stems with a pin to help them stand taller longer.

Don't forget to change the water every day.

> **Sticky Note:** If you want to send flowers to a friend who lives in another state, call the concierge at a cool hotel in the area (like the perpetually hip nationwide chain of W Hotels) and get florist recommendations.

Your walls are bare and you need to choose a piece of art, but you don't know where to start.

Andy Warhol said, "An artist is somebody who produces things that people don't need to have."

Warhol, as usual, was likely trying to provoke, but he's right—you don't need art, you want it. Follow your instincts when it comes to selecting objects you're going to live with over a long period of time. You want art to inspire and challenge you, but not to jar you—or, heaven forbid, scare you. I have a friend who dropped a ton of cash on a disturbing work by a brilliant and famous artist, and while it was an "important" piece, she couldn't bear to look at it. She would've been better off spending $20 on a reproduction of a 1940s movie poster, because now her investment is wrapped in brown paper in her garage.

Here's what I think about when I'm in the art market:

- It doesn't matter if a piece matches your decorating scheme. In fact, a painting that coordinates with the drapes—and the sofas, and the rug—looks matchy-matchy and cheap.

- Don't buy art because it's a good investment. A painting or sculpture should be something you're dying to look at. If you're just looking for something to do with your money, put it in a savings account.

- Don't be afraid to turn everyday objects into art. Pose your collection of sock monkeys on a glossy floating shelf; cover a whole wall in Polaroids; frame your favorite piece of vintage clothing in a shadow box. Remember, the artist Marcel Duchamp was a mean painter, but it wasn't until he stuck a urinal in an art gallery that he earned his spot in the Western canon.

Sticky Note: Always avoid hanging a picture in direct sunlight. Doing so will fade its original colors, diminishing its beauty (and value).

You can't hang a picture straight.

First, find the wall stud. If you try to suspend a picture from drywall, it may well hang straight—before it crashes tragically to the floor. To locate the right spot for your picture hook, get an electronic stud finder. No, this is not a hottie homing device—it's a little gadget that measures the density of a wall and lights up when you've hit a good spot for a nail.

Then decide where you want the art to hang. Aim to position the center of the picture at the average person's eye level, which is between 5'5" and 6'2". Use a pencil to mark this spot on the wall.

Grasp the wire at its midpoint and pull it toward the top of the frame, as taut as possible. This is where the nail will be when the picture hangs. On the back of the frame, measure from its lowest point to the top edge. Take this measurement and use it to draw another spot on the wall, beneath the first one. Nail a picture hook into this second spot, hang the picture, and adjust the frame so that it's centered. Check its placement with a level by placing the level on the top of the frame. Once the bubble is centered, the level is perfectly horizontal. Voilà! Straight-up awesome art.

More on the art of hanging art:

- Before you hang pictures, make templates by tracing around each frame on a piece of newspaper and cutting them out. Arrange the templates on the wall (using removable tape) until you are satisfied with the composition.

- In a grouping of pictures, place the strongest piece in size, color, or subject matter first, and let the others fall into place around it.

- Create alignments, so the viewer's eye has lines to follow. These visual lines may be horizontal or vertical. If a wall contains many pictures, there may be several of these alignments. Any two frames should have a common line, horizontally or vertically.

Sticky Note: To prevent the visual clutter that can come with a wall of random family photos, make black-and-white photocopies of color snapshots and intersperse them with vintage prints. Pull all of the images together with white mats and frames.

DECORATE TO IMPRESS

Your Parents

Objective: To convey that you are mature, tidy, and ready to serve as caretaker for Grandma's killer Swedish modern coffee table.

Solution: Keep things fresh and light. Eighty-six anything that suggests either a college dorm room (Christmas tree lights, unframed posters) or their little princess' old bedroom (frilly lamps and lacy linens). Choose a simple but sophisticated color scheme, like classic white with black accents. And clean, clean, clean. For Mom and Dad, a spotless apartment is going to say more than a way-out decor. Oh, and—needless to say—hide the condoms.

Your Boyfriend

Objective: To convey that you are a domestic goddess, without being so girly that it freaks him out.

Solution: Think comfortable but not fussy, sexy yet approachable—like the aesthetic of a W hotel room. Your couch (and bed) should be soft but not crammed with extra pillows and throws. Choose functional furniture (no pretty antique chairs that break on contact) in basic hues. That said, never let him forget you're a girl. High-thread-count sheets and easy-to-clean luxury textiles like cotton velvet will keep him coming back for more.

Your Future Mother-in-Law

Objective: To convey that you have your life together and will provide a wonderful home for her little darling.

Solution: Stash away anything that reflects your former single life, like hilarious spring break photos or your French erotica collection. Instead, prop up baby pictures of you and her little one next to vases of fresh flowers. Let her know that you've got your eye on your future with basic and well-built pieces of furniture (no futons or foam). And don't neglect the Swiffer—a spotless apartment will have you on first-name terms in no time.

Your Fancy New Girl Crush

Objective: To convey that you're the fabulous BFF she's been waiting for.

Solution: Bold colors and crafty touches will show off your flawless taste—try picking up original art and mod screen-print pillow shams from Etsy.com. And she's bound to appreciate your thrift scores, so paint a vintage table and chairs in cobalt blue or aqua to brighten up a dull kitchen. If all else fails, serve a bunch of fancy cocktails and salty snacks.

Sticky Note: It's no longer necessary to spend thousands if you want your home to look like it's been decorated by a pro. Many top brands are doing secondary lines for chain stores—like Smith and Hawken, Victoria Hagan, Dwell Studio, and John Derian for Target. There are also a crop of sample sale Web sites, like One Kings Lane, which sell fancy furnishings at crazy discounts.

If You're Missing the Gene for Organization

A few years ago, I was both humiliated and flattered when a colleague gave me a book about clutter. Clearly, she was telling me she had noticed my piles, but since the book was about clutter being a sign of genius, she must've thought she was paying me a compliment. While I was glad to be in the company of smarty-pants like Einstein and Alexander Fleming—who discovered penicillin by neglecting to wash a dish on his desk—these men were not known for their aesthetic sensibilities. Being a Modern Girl with a media career—not a middle-aged man toiling away in a laboratory—I decided it was time to stop living like a pack rat. Here are some tips that helped me streamline my space.

First, though, go through your stuff. Before you can organize, you must ruthlessly assess your things and toss or donate the junk you don't use. Make piles—you like piles, right? Separate the things you really love and/or need from the things you don't. Donate everything in the second pile to your local thrift store or put it up for sale on eBay or Craigslist.

On Your Desk

Corral the small stuff, like with like, into a series of matching baskets or pretty storage containers with lids. They sell fantastic clear plastic ones at Target and the Container Store. This way, you'll know exactly where to go to find what you need, but you can still hide it to create the illusion of neatness.

Use cardboard magazine boxes to organize files you need to access frequently. Keep the open end facing you and the closed end facing out. You can also keep an open file box by your feet to store files you access frequently. This will save you from making constant backbreaking trips to the file cabinet.

Buy a silverware tray for your desk drawer to organize pens and paper clips.

Keep a chairside table, like a bedside table, to hold items you need close at hand but don't want to display on your desk.

Cover your entire desk with a large clear blotter. This can eliminate the need for a separate mouse pad, since it provides traction, as well as make it easy to clean up any spills. Plus, you can replace it much more easily than you can your desk.

Clean your monitor, mouse, and keyboard with a slightly dampened Mr. Clean Magic Eraser. You'll be amazed at how much grime it picks up!

Use a clear acrylic cookbook holder to keep books and documents open next to your monitor, so you don't need to keep looking up and down as you type.

Hang photos instead of displaying them on your desk. You never want to confuse memories with clutter.

In Your Pantry

Create a system. Divide shelves into five sections and label each one to ensure your food will stay organized regardless of who's making dinner. Create one section each for baking supplies (flour, sugar), starches (pasta, rice), breakfast items (cereals), canned goods (soups and sauces), and snacks (chips, crackers). It's not rocket science, but the key is to minimize the amount of movement you need to do to get the ingredients you need.

Display pantry staples in clear glass jars, so you will know when you are getting low on sugar or fusilli. Many Modern Girls love rectangular jars in lieu of circular ones—they maximize space.

Large wire baskets prevent large items like paper-towel rolls or bottles from toppling over.

Double your shelving with stainless-steel platforms that maximize vertical space. They also make great storage for kitchen linens.

Put the stuff you use most frequently close at hand—on the front of the middle shelf.

Store cookies and other snacks in tilted wide-mouth jars at eye level—so you can snag a cookie on the fly, and keep little hands away from treats. If you're watching your sugar intake, store them on top of the refrigerator so you'll need to be shamed onto your tippy-toes.

Attach metal clips from an office supply store to the back of the pantry door to organize takeout menus and spare recipes.

Instead of keeping your knives in a drawer, hang a large magnetic strip near your drainboard or chopping board and store knives there. They'll stay sharp and take up no space.

At Long Last

A guide to how long those staples are really good enough to eat—and when you should toss them and bring in the new!

Item	After opening, lasts for
Baking powder, baking soda	6 months
Bouillon cubes	2 years
Cocoa mix	6 months
Cocoa powder	2 years
Coffee beans	2 weeks
Coffee, instant	2 months
Cornmeal	1 year
Flour, all-purpose	8 months
Honey	1 year
Pancake mix	9 months
Peanut butter	3 months
Spices, ground	2 years
Spices, whole	4 years

In Your Bathroom

Aim to leave nothing on your countertops. Use shelves, drawers, and closets first. Invest in drawer organizers to make the most out of your space. For a superaffordable option, try plastic or chrome silverware trays.

Try installing wall-mounted dispensers for shampoo, conditioner, and shower gel to get rid of bottles. This will enable you to buy inexpensive bulk brands since you don't have to worry about packaging. I like the ones from simplehuman the best.

As counterintuitive as it may sound, never keep your medicines in your medicine cabinet. Privacy issues aside, the humidity in the bathroom isn't good for them. Line them up on a pretty lacquer tray and keep them on a low shelf of your bedside table, or in the kitchen cabinet near the water glasses.

Store toilet paper in a tall glass cylinder. It looks chic (as chic as toilet paper can possibly look, anyway), and your guests will never have to ask you to slip a few squares under the door.

In Your Linen Closet

Store linens by set, not by sheet type.

Organize linens by season—flannels with flannels, etc.

Don't cram your sheets into shelves: air needs

to be able to pass over everything in order to keep it smelling fresh.

Put valuable linens on top shelves—you don't use them often, and they'll be safest from water damage up there.

Hang tablecloths on a hanger, folded lengthwise.

Now that you've got a sense of how to make your stuff at home in your home, read on to see how to best cope with the people living there.

Home: Living There

When my now-husband Marcus and I moved from New York to California, we bought the first house we saw. Having just left behind a tiny one-bedroom apartment, we were amazed at the amount of space before us when we crossed the threshold into what seemed like real estate nirvana. Never mind that it had a bizarre layout, no closet space, and a climate control system from the previous century—this house seemed like a dream. About a week after we moved in, we realized we'd made a terrible deal. Thinking back, we realized that the real estate agent had seemed a little too overjoyed when we jumped to sign the contract. She had probably been praying to the mortgage gods, "Send me a New Yorker. Send me a New Yorker." Put them anywhere nicer than the F train platform in the dead of winter, and a New Yorker is grateful.

Still, we made that overpriced house a home, and looking back, I wouldn't have it any other way. Here's how to make the most of your home without getting lost in space (or lack thereof).

You're moving in with your boyfriend, and you have opposing aesthetic sensibilities—and twice as much stuff as you need.

The two of you have finally bitten the bullet, split the rent, and shacked up. Now, it's time to make some compromises. Who's got the most comfortable mattress? If you're nervous about shedding your pillow-top, consider renting a storage space for your duplicate furniture (and mix tapes from ex-boyfriends). That should take some of the urgency away from arguing over his sectional versus your chaise longue. When it comes to colors, try choosing paint swatches separately and then coming together to see if there are any overlaps (hues that don't skew masculine or feminine include muted shades of blues, greens, and grays). For linens, skip the patterns—florals, plaids, *Star Wars*?—and get plain sheets with a good thread count. (BTW, consumer studies show that anything higher than 300 is meaningless.) And, finally, art—consider making a salon wall with a combination of both your framed goodies. His French *Planet of the Apes* poster will have new resonance when framed and hung alongside your signed picture of Rosalind Russell. Just don't forget the perfect accompaniment—a portrait of the two of you.

You're sharing the tiniest living space (or, worse yet, bathroom) with the biggest of personalities.

When having to live in the confines of a small apartment, it's best to make clear guidelines with your roommate or lover. Designate storage space for each member of the household and be scrupulous about staying within the limits. Figure out a bathroom schedule. Do you have a tight schedule in the morning? Knowing this may encourage your roommate to straighten her hair in her bedroom instead of locking herself into the WC. And try to be sensitive about noise—if you like to watch TV into the wee hours, consider buying wireless headphones. You won't miss Jimmy Fallon, and your boyfriend can still get some shut-eye.

You have noisy neighbors—or you are the noisy neighbor (according to the people next door, hmpf).

Regardless of whether you're the victim or the perpetrator in this situation, here's what you need to know: When it comes to neighbors, the general rule is that if she can hear your iTunes at her place (or you can hear hers in yours), the music is too loud. (This may account for the enduring popularity of prewar apartment buildings, where the walls are thicker and thus fewer battles of the bands erupt.) If you're the perpetrator, consider speaking with your complaining neighbor to get an idea of her daily schedule. That way, if you simply must rock out, you can let it rip when she's away. If you're the victim in this situation, the first thing to do is to speak with your neighbor in a clear, calm, non-confrontational manner about the issues that are affecting your inability to enjoy a quiet peaceful home life. If this doesn't help, the next thing to do is slip a nice handwritten note with something that acknowledges you've already spoken and clearly states your issue. Here is an example:

Dear XX,

Per our conversation last week, I would appreciate your consideration of my effort to enjoy a peaceful home life, and while I know it is unintentional, the volume of your music/sex/parties is preventing this from happening. Please call me if you would like to discuss further, and thank you in advance.

Sincerely,
Jane

If this doesn't work, the next step is to go to your landlord/manager. Let him know that you've tried to resolve the issue politely without his assistance but that the problem continues. Tell him that he needs to step in as mediator and property owner to figure out a solution that works for both of you. And if that doesn't work and your neighbor is still having debauched weeknight house parties, call the cops and resubmit your complaint to your landlord. If you've executed all of the above and nothing works, your only other next step is to move out. While you could force the landlord to take action in court, it is probably not worth it unless he won't give the security deposit back. Your neighbor is not going to change, and your landlord is a slumlord in all senses of the word.

The hall of your apartment building smells like a Green Day concert, circa 2001.

First, check out your state's law for smoking in apartment buildings. California is ahead of the curve on this, and the right to a smoke-free apartment is currently percolating through the courts. What would this mean? In California, if your neighbor is a chain smoker and you like to keep your windows open when it's nice out, then it would be illegal to smoke nearby, as there is a law that protects residents from drifting secondhand smoke in multiunit residential buildings. For more information, check out smokefreeapartments.org/general_03.html.

Your heating bill is astronomical—and you're still freezing!

Close your drapes at night and open them in the morning. Use the sun to warm the house, then trap the heat inside, and you could save up to $200 annually.

Don't shut interior doors. Inhibiting airflow from one room to another makes many heating systems work harder.

Inspect your furnace filter—if you have forced air, a clean filter will save up to $100 a year. The filter can be hard to find (it's usually behind a panel near the blower) but is easy to change. Check it every three months. If you can't see light through it, get a new one at the hardware store for about $3.

Seal your windows. With $50 worth of vinyl weatherstripping and a little patience, you can snug up five windows and save as much as $100 a year.

Have your furnace serviced. For $75–$150, hire your oil provider (check to see if the service is part of the contract), furnace installer, or other professional inspector to look for leaks, worn parts, and other inefficiencies that may be costing you a few hundred dollars annually.

Order a home energy audit. Get a list of fixes specific to your home, at a cost of about $400. For auditors in your area, visit EnergyStar.gov. Your utility company may even offer a rebate.

Heat yourself, not the house. Lower the thermostat every time you leave your home and before you go to bed and you could shave up to 5 percent off your energy bill.

You want to go green, but you don't have the cash.

If you think being green requires having lots of green, you're living in the nineties. Sure, the cost of organic produce and natural cleaning products can really add up. But while a lot of us think about going green in terms of buying stuff, it's really more a matter of not buying stuff. You don't need to buy a Prius; just take the bus once in a while. You don't need to install solar panels; just unplug your appliances when you're not using them. You don't need to buy a new energy-efficient dryer; just try a clothesline. Pretend you're Laura Ingalls Wilder, living in a little house on the prairie. Living old-school doesn't cost much.

Think of simple things, like:

- Using a locally raised, free-range whole chicken to make a roast chicken dinner, next-day chicken salad, and homemade soup—instead of buying a preroasted chicken, a chicken salad sandwich at the deli, and a package of ramen.

- Using staples like white vinegar, baking soda, salt, and lemon juice to clean your home.

- Investing in a water filtration system and an aluminum bottle instead of buying bottled water.
- Joining an organic-produce delivery co-op.
- Starting a veggie garden in your yard or on your fire escape.

You want to go green, but burlap isn't really your style.

Being eco-friendly is no longer about hugging trees, wearing Birkenstocks, and eating bark. These days it's not only trendy but tremendously important that we all do the most we can to start helping this planet. Now that doesn't mean I expect you to give up all of your favorite products, jet travel, and Diet Coke (goodness knows I didn't). But you can, as *Gorgeously Green* author Sophie Uliano says, choose which shade of green you want to be. I prefer a sort of Kelly green. I do what I can, but it still reminds me of Hermès.

I get some of my best advice from Ecofabulous.com, Ecostiletto.com, Idealbyte.com, Healthybitch.com, Vitaljuicedaily.com, and Thedailygreen.com. All of these are packed full of great info on ways to be a little greener but still glamorous. And for those of you just moving from sludge to green, here are a few easy ideas (and go to greentipsblog.com for some more):

- Invest in a reusable bamboo utensil travel set and to cut down on the use of plasticware when not at home.

- Use organic, natural cleaners (or a homemade solution of baking soda) and vinegar instead of possibly more toxic ones.

- Try to use major appliances during off peak hours to save energy and money!

- Use reusable bags! I like the ones at EcoGreenBags.com.

- Keep a cactus nearby in your home and your office to freshen the air as you work.

- Save trees and reduce junk mail. Go to Catalogchoice.org to take yourself off of unneeded lists.

- When bringing your lunch to work, use Tupperware instead of plastic bags. They've got some supercool designs and you'll save money.

- Request e-statements for bills instead of paper statements.

Sticky Note: If greening your home seems unmanageable, start small by greening your beauty routine. Replace active chemical ingredients in skin care with natural ones. Soy, for example, is superhealthy for your skin—it's a natural source of antioxidants like vitamin E and an anti-inflammatory. It helps calm skin, improving its tone, texture, and radiance. It's healthier and greener to choose these natural skin soothers over a synthetic product containing potentially harmful ingredients like petrolatum, which is used in many moisturizing beauty products. You don't need to break the bank, either—try Jergens Naturals Moisturizer, a 97-percent natural, paraben-free formula containing soy extracts to help soothe and promote healthy skin. Target now carries numerous organic cosmetic lines—luxury at a great price. I love Weleda.

PART 4

Sticky Situations in Your Family

Where is this mythical "normal family" we all spend our childhoods wanting to be a part of? Like the abominable snowman or Snuffleupagus, no one is really sure it exists. In fact, according to a recent study, seven out of ten American youths are living in nontraditional families.

Traditional or not, family dynamics can be hard to navigate. As clichéd as it may sound, most family problems are a result of poor communica tion. For example, I have a friend who spent her entire childhood thinking her stepfather hated her, so she ignored him. When she was twenty-one, he revealed he'd always thought she hated him, so he had been giving her space. Now they're BFFs, but they wasted ten years!

Me? I'm pretty typical for someone of my generation. My parents split up when I was five, my mom went back to work, and I became the woman of the household. I could organize a pantry and cook a mean TV dinner, all while getting my homework done. Today, my family looks as stereotyped as it could get—a great hubby, a boy, and a girl, and we even got a golden retriever to fit the bill—but trust me, we are far from "normal"!

If there's anything I've learned from my highly unscientific study of family relationships, it's that it's better to be honest than quiet. Here are a few ways the truth will set you free.

You feel like you're living in a scene from *Reality Bites*. Every time they see you, your parents can't help expressing how much they think you're wasting your life. And every so often you begin to wonder . . .

How old are you? How educated are you? How financially and emotionally independent are you? What goals and aspirations do you have for yourself, and what is your plan for achieving them? How do you spend the actual hours of your days (and nights)? Do you have a job, and if so, do you enjoy your work? Where do you see yourself going in the next five years? Ten? Do your friends and/or social set reflect your values, ambitions, talents, and habits? Do you abuse drugs or alcohol? Are you a reliable person others can count on? A social butterfly and go-getter or a quiet person who likes to spend your spare time reading? The point is, people within a family can have different values when it comes to relationship and career choices. The parents of one guy I know told him he was wasting his life when he announced he had decided to become a priest. The important thing is to be your own person, someone who is bolstered and motivated by her own thoughts, beliefs, and feelings. If you're happy, healthy, productive, and passionate about your life and what you're doing, you shouldn't let anyone make you feel as if you're on the wrong path.

All that said, you must ask yourself the obvious: Are you wasting your life? If deep down you know your parents are right, don't be afraid to ask them for help and guidance (not the financial kind!). Everyone needs a little boost at one time or another.

One of your parents has done something you can't forgive.

Ingrid Bergman supposedly said that forgiving is easy; it's forgetting that's impossible. So, first, what is the parental transgression? If it involves violence or violation, the breach of a personal boundary that desecrated something important, you should seek professional counseling. There are things we should not attempt to shoulder or bear without the intervention of an advocate. If the misdeed was less grave but nonetheless significant enough that you find yourself stuck in blame and condemnation, why not talk to your parent and tell him or her how you feel? Or write a letter, which can be read privately and then responded to. To be disappointed by a parent is devastating, but at the same time, it carries the benefit of allowing us to see, and accept, our parents as human. Listen, I'm neither judge nor jury here, but I will tell you that in the last conversation I had with my mom I was sort of bitchy for no reason. I was twenty-one, and she died unexpectedly that night. When you're young it's rather fun to be melodramatic. If your parents did something awful, they deserve the silent treatment. But if they didn't let you get yours ear pierced until you were sixteen, you're going to wish you had behaved differently. Trust me.

You need to tell your family you're gay . . . and they aren't as accepting as you would like.

If you're out and proud and your parents are still trying to fix you up with every able male outside the family circle, it's really time to sit them down for a discussion. Explain that you love them and you want them to be part of every aspect of your life. It could be that they are having a hard time because they want you to have a conventional, happy life with no difficulties to surmount. Tell them that for you, happiness can only come with being open about your sexuality. Now, if your family has religious beliefs that are anti-homosexuality, you may have to speak to them in their own language. Suggest that they pray for you—and accept you. Say, "I know you don't understand why I live my life the way I do, but I'm asking you to love me as I am."

Help them understand that this isn't a "choice" or something you are "doing to them." In fact, try to help them realize that, unlike when you dyed your hair pink at fourteen and got your belly button pierced, this has nothing to do with them. Try to remind them of all the times they told you they just wanted you to be happy. If they truly can't accept you, let them know you all might need to take a little break from each other until they are willing to accept you as you are.

If, on the other hand, you're straight and your parents are convinced you're a lesbian, you've got another problem. Humor is a great way to approach it. The next time they drop pointed references to loving the *Ellen* show or Rosie O'Donnell's made-for-TV movies, say, "Look, I hate to disappoint you, but I'm straight." Let them know that you're happy that they would accept and love you no matter what your sexual orientation.

Your in-laws think you were a bad choice; your parents think he's a loser. Basically, no one is happy but you two.

Ignore both families—the only opinions that matter are those of you and your spouse. Don't take your parents' and in-laws' disapproval personally. It's not about you; it's about them and the fantasies they had about the perfect mate for their child—the groom that would turn you into Angelina, the bride that would turn him into Brad. Don't internalize the judgments or expectations of others, and don't let them come between you and your mate. For your marriage to succeed, you have to be on each other's side. The walls around a relationship should be like Teflon, strong and with a surface that keeps rotten eggs and tomatoes from adhering. Of course that doesn't mean you shouldn't *try* to prove your parents wrong. Show them how in love you are. Help them see that the match is right by killing them with kindness. Continue to invite them over, attend holiday events, and be the model of the perfect daughter-in-law.

However, there is an exception here. Did you two meet, fall in love, and decide to get married all within the course of, oh, say, two weeks? If so, strike the above and take a minute to listen to your parents. They just want to make sure you are making sensible choices. If you're rushing marriage because you fear that by taking time you may slip away from each other, *you need more time*! If, however, you truly have had time to be together and get know each other, have faith in each other. That's what Marcus and I did, and fourteen years later we've proved even the most skeptical person wrong.

Your parents are divorced, but you want them both to come to your wedding.

Of course you do! And why not? It's your happy day, they're your mother and father and you seek their

Dear Jane,

My sister wants to bring her new boyfriend on a girls' vacation we've been planning all year. I was really looking forward to some sister time—not to mention the fact that he's an unemployed dud who grunts at us. Can I ask her to come by herself?

It's a sad fact that every woman is going to date a loser sometime. By telling your sister what you think of her man, you risk alienating her and damaging your relationship. But if this is a trip on which you'll be spending your hard-earned cash, then by all means ask her to leave Mr. Wonderful behind—unless he's willing to pay for the both of you. Explain that you were looking for a different kind of trip, and if she wants to go on a trip with her boy, then you might have to bail. While you aren't meaning to make her choose, this isn't the threesome you were hoping for.

If you do end up on holiday with the happy couple and your opinion of the male half doesn't change, pull your sister aside and ask her—in the blessings on a union that you hope and pray will work out better than theirs did. There are whole genres of film and fiction, not to mention magazine, newspaper, and online advice columns, devoted to the subject of avoiding murder and mayhem when nuptials include divorce among the blood relations. There's no overall panacea here, because the chances for success depend upon your own specifics—how long your parents have been divorced, what their relationship is now, who's still nursing what hurts or grudges, who tends to revisit the past after too many flutes of Veuve Clicquot.

Begin by being honest and direct. Approach each parent and tell them what you want: that your heart's desire is to have them both at your side on this special day. Hear what they say, with this caveat: don't become the negotiator or messenger, running between one and the other with concessions and demands. Your wedding day is about you, not them. Be clear, and don't allow them to use this day as a battlefield for their old issues. Having said that, if you haven't seen your dad for twenty-nine years and your mother raised you and put you through private school and law school, don't be surprised if she's a bit hurt if you want him to walk you down the aisle. Traditions may be traditions, but in this new world of divorce it's anyone's game. Try to talk to each of your parents separately about what you want from them at the wedding and see if there are things you can achieve together. And, by the way, everyone in attendance is expected to

kindest, most nonconfrontational way possible—what she loves about him. Either you'll find out some little-known positive points about her boyfriend or she might find herself tongue-tied and start rethinking the relationship. Either way, you'll have made an effort to understand her.

act like an adult except the flower girl and ring bearer. If one of your parents can't be the bigger person, limit his or her role to a small one and remind them that you want your marriage to end up far differently than theirs did.

Sticky Note: Tell people the baby's gender in advance if you want, but not the name. You'll never get a universal favorite!

Your family expects you to name your baby after a particular family member, but you don't want to.

Fact of life: Your family will always have expectations of you that you are disinclined to fulfill. Naming a baby falls squarely in the realm of your business, not your mother's or your father's or your sister's or . . . you get the picture. As in many family matters, what's important here is not the message of your independence but how you deliver it—neutrally is best, and with a smile. Don't be disingenuous or deceive anyone into thinking you'll do something you know up front you won't, but, on the other hand, many family messes can be averted by graceful sidestepping, as in, "Thank you for that suggestion. I adored Aunt Hortense and would love to do her the honor. But Harry and I have been deluged with advice on names, and we're sorting through the whole blessed assortment, plus trying to agree on our own personal favorites." Don't dismiss your family's ideas, but don't let them impose them on yours either. A compromise? Consider using the first letter of the name in question. And don't announce the name until it's a done deal on the birth certificate. Naming is not a democracy.

You've heard of rich man/poor man, right? Well, what about rich sibling/ poor sibling? Either way, it's sticky.

So what? Different strokes for different siblings, different vocations at different rates of remuneration. Yes, your investment banker sister's income and net worth will be significantly more than yours if you're a high school art teacher. This is not important. What's important is, are you happy and fulfilled in your life? Motivated? Challenged? Happy to get out of bed every morning and go about the business of your day? Don't get me wrong, money is a welcome and useful commodity in life. It can confer a certain ease and status, and, like a sibling's having beautiful looks or extraordinary athletic ability, it is a manifestation of inequity—but only a surface inequity, not a core currency of value and worth. And by the way, life is long. Your quirky sister's passion for yoga may turn into a burgeoning business that makes her more money than your 9:00-to-9:00 job in advertising. And as my brother once prophetically said to me, "What makes him happy doesn't make me happy." So very true. (Though why a stroll through the first floor of Barney's doesn't put a smile on everyone's face is still a mystery to me.)

Everybody wants to go on a big family vacation. Everyone, that is, but you.

It's a simple fact: family bonds are worth caring for and tending to. It's a week; it's a vacation. Suck it up. Sacrifice or postpone your personal preferences for the sake of the whole. If your idea of bliss is a week with your mate in Paris but everyone else wants to cavort on the beaches of Hawaii or, worst-case scenario, rent an entire floor of a hotel at Disneyland, go with the flow. Get thee to Honolulu, or Anaheim, with a smile on your face and a big fat *om* resonating in your throat. You'll feel so big-hearted and virtuous that upon your return home, you'll book Air France as your reward.

You feel you're always being super-generous to everyone in your clan, while no one is generous to you.

In some ways, giving too much in a collective relationship is like giving too much in a one-on-one. When it's you 99 percent of the time and you feel angry, resentful, sucked dry, and cheated, something is out of whack and, I'm sorry to say, your loving generosity is neither loving nor generous. It's like overwatering a plant. You're not sure how much H_2O it requires and your instincts tell you to nurture it, so you repeatedly soak the soil and end up killing what you thought you were helping to flourish. In general, the give-and-take of any relationship shouldn't feel like you have to keep score, and if it does, pull back and see what happens when you give your giving a rest. You may have a lot to give, and you may want to give it, but don't set the bar so high that there's no room for future growth or mutuality. Sometimes doing less opens up the space for others to do more.

Your mother-in-law—or another member of your husband's clan—overhears you talking about her.

This is one of those times when you have to be a straight shooter. No pussy-footing or hand-wringing or insincere, mealymouthed, girly retractions. Woman up—swallow hard, look whoever happened to be in earshot straight in the eye, and in the most level voice you can manage say something akin to "I'm so sorry. It wasn't my intention that you hear that."

What comes next depends on how the person responds. If what you said wasn't too harsh or inflammatory and if the other person isn't insulted or infuriated, perhaps you can restate it in a kinder, gentler manner: "Forgive me. I should have taken this up directly with you earlier. What I meant to convey is . . ." Even though you might be embarrassed, don't forget you're the perpetrator here. It is up to you to take responsibility and defuse the situation, and the best way to do this is to come off as sincerely contrite and apologetic rather than defensive or argumentative. As my grandmother used to say, if you step in something stinky, don't keep wiping your foot on the mat.

You're beginning to feel like you're living in a Disney fairy tale, only without the handsome prince: your stepmother is evil and undermining.

In literature and film, the word *wicked* or *evil* so often precedes *stepmother* that we don't even think about it. Snow White, Cinderella—we all know the stereotype: manipulative, jealous, catty, self-centered. The one who casts her stepdaughter aside in favor of her biological girls, or sends a woodsman out to kill the stepdaughter because she's the fairest in the land. To be presented with a stepmother who is sympathetic and supportive, as in *Juno,* is a radical concept.

Where does this big red letter *S* come from? Let's begin with acknowledging that stepfamilies are hard work. A lifetime of work. Studies have shown that stepmothers have the most difficult position in a stepfamily system. They come to a new marriage, some for the first time, some without children of their own, and they try to make a new family. Often, stepmothers end up feeling taken for granted by their husband and stepchildren for the work they do. They feel like they're raising the child without reciprocal affection and acceptance. If the biological mother is unhappy about the stepmother's involvement with the child, this adds another component of misunderstanding and lack of appreciation. There are expectations all around that she jump in and love the child as her own, but the reality is that sometimes stepparents and stepchildren don't even like each other.

All of which goes to say that an "evil and undermining" stepmother may just be someone who is trying to figure out how to maneuver between you, your father, and your biological mother. Or she may have no experience with children and unrealistic ideas about what you should look like and how you should behave. It may be difficult for her to accept that there was another, first family that came before her and she'll never be part of that history. So first ask yourself if it's possible that your stepmother's true intentions are good. If the answer is yes, whether you like her or not, she is not wicked and you should find a way to co-exist. If the answer is no—if she is nasty or cruel to you outside your father's presence, if she goes out of her way to make you feel worthless and unimportant, if she threatens you or is physically abusive—talk to someone, hopefully your father, but if he isn't available or receptive, find a neutral party who knows your family situation.

No matter how mean your stepmother seems, be nice to her. Then she'll have no excuse for her behavior. If she keeps being an evil stepmother while you're being nice, collect evidence. Write down the meanest things she says to you, but only if they were uncalled-for. See if you can tape-record or videotape the two of you alone. Use this to show your father how she really treats you. But do not lie to (or about) her, and do not push her buttons to make her look bad.

Ask your father for one-on-one time once per week so you can maintain your own private relationship with him and tell him things that are on your mind—your stepmother being one, but not all, of your concerns.

Do things that are fun for yourself—classes, hobbies, activities outside the family. Disengaging from struggles instead of letting them consume you is a healthy move. Concentrating on things that make you happy gives you more energy for the things that are tough.

You're fighting with your siblings or stepsiblings over an inheritance.

Fighting over an estate (even if small) is a particularly stressful and destructive display of human nature. Old sibling rivalries and feuds erupt, greed rears its ugly head, spouses of the heirs join the fray, families get split apart for life, and in the end no one really wins except the lawyers you've hired to contest the will or asset distribution. And the saddest, most ironic thing of all is, your parents' inheritance was—hopefully—their way of trying to help you live a better life.

Perhaps the first thing to ask yourself is: Is this worth it? Even if you feel the terms of a will are inequitable, disproportionate, and unfair, what is more important to you, harmonious family relationships or money and possessions? Siblings who don't feel close at the time of an estate settlement often grow closer over time. They forge a new, reconfigured family structure around their absent parent or parents. Death is a double-edged sword. It deprives you of the family frame you have known since birth, but it also gives you the opportunity to start fresh, put aside old roles and antagonisms, and adopt new ways of relating to one another. A bitter fight or drawn-out controversy can easily destroy this possibility before it's had the chance to begin to flower.

Question number two: What are the alternatives? Is there a way you can bring people together? Avert a long, drawn-out battle and maybe a lifetime of alienation? Even if yours is a family where harmony doesn't take place naturally (and most families are like this to one degree or another), heirs can make a determined effort to reach an amicable settlement. Auctions and round-robin choose-ups are a good way to divide household items. When sentimental value comes into play, such as sisters arguing over their mother's wedding ring or other jewelry, consider a sharing plan where each party keeps it for a year and then passes it along. In many families, items that were purchased for a parent go back to the original giver upon the parent's death. Special consideration should always be given to the sibling who has been the most consistent caregiver to the parent who has passed. Using a "who needs it most" principle can also be helpful, such as giving your parent's car to the sibling whose ancient Chevy just broke down.

Divorce, remarriage, and the resulting extended and blended families create an even greater opportunity for greedy in-laws and stepsiblings to be at each other's throats. One thing to remember is that, with very few exceptions, stepchildren, stepsiblings, and stepparents of the deceased have no inheritance rights in the absence of a will. In most states, if someone dies intestate (without a will), the surviving spouse, children, parents, siblings, nieces and nephews, and other next of kin inherit, in that order.

Your Kids

One of the craziest things about becoming an adult is shifting from being a kid to being a parent. I keep waiting for the magic moment when I'm going to feel my age and instantly know the right thing to do—will there be a magic wand involved?—but that wish in itself is childish, I guess.

You recently had a baby, and reading this book counts as the most adult conversation you've had in weeks.

One of the hardest things about parenting, especially the first time, is feeling as if you are alone in your struggles. What you need are moms to talk to who are going through what you are. So, put down this book, brush your hair, nestle that baby in a sling, and head out to the local organic store or park to find others like you who are immersed in the baby world. Sound like too much work? Okay, fine—maybe tomorrow. Anyway, sometimes real-life moms don't want to share the "real-life" problems, anyway. Even the best of friends sometimes sugarcoat their parenting ("My baby slept through the night at two months" really means "My baby slept from 12:30 A.M. to 5 A.M."—only a decent night's sleep if you're accustomed to taking the red-eye). So today, log on to one of the amazing blogs about parenting that speaks to you, not *down* to you. Some of my favorites are Dooce.com, Busymom.com, Musingsofahousewife.com, Modernmom.com, Mommytrackd.com, Thecradle.com, Truemomconfessions.com, and Cafemom.com. And if you didn't join Babycenter.com when you got pregnant, do so right now—this site sends the best weekly e-mails to track your baby and toddler's progress. It's like a developmental horoscope that's eerily accurate.

In addition to these blogs, more and more local communities have sprung up, with great newsletters like Jenslist.com in Los Angeles and Westvillagemoms.com in New York City, which give great local recommendations. If you want to join a dialogue versus just reading one, try one of the great mom groups on Twitter (@mombloggersclub and @coolmompicks), or communities on Bigtent.com and Ning.com, which allow you to talk right back, too!

You actually do have a favorite child (and no, Jack and Lilia, *I do not*).

You're not alone. In a recent poll by the British Web site Netmums, 16 percent of respondents confessed to having stronger feelings for one particular child, while more than 50 percent said they loved their children equally but in different ways. To truly love one child more than another would be an awful source of guilt and shame, but few of us really feel this way. If one of your kids smiles and chats through family dinner while the other wears his earbuds at the table and pushes food around his plate, it's natural to prefer the company of the first child. Preference that has nothing to do with behavior, though, is something that warrants a family trip to the shrink. But, crazy as it may seem, a 2008 Temple University study found that favored children, in adulthood, were less satisfied with their lives than their siblings who did not perceive themselves to be the favorite. So that should make you feel a teensy bit less guilty about sometimes giving Johnny the bigger piece of cake. If you don't have a favorite (like me) but your kids think you do, when they are old enough to understand sarcasm, try this: "Yes, you're right—I *do* love your brother better." It defuses the subject quickly!

Your mate and kids are always ganging up on you.

Eleanor Roosevelt famously said that no one can make you feel inferior without your consent. The next time you feel ganged up on, raise your hands and call a time-out. Explain that you expect your opinions and your authority to be respected. And if the perpetrators can't seem to grasp and accept this en masse, talk to them one by one. Revert to Julius Caesar's dictum to divide and conquer, which worked exceedingly well in the Gallic Wars.

You woke up and realized your kids are just plain spoiled.

Many experts consider overindulgence to be a form of child abuse. By spoiling your kids at every turn, you give them a false sense of how the world works and set them up for a lifetime of disappointment. In the real world, we don't always get what we want. Children need to be equipped with the skills and resources to help them meet the challenges of independent adult life, and they won't learn these skills by being fed Cold Stone Creamery ice cream while they play Wii. If you have no idea how to say no, don't kid yourself: providing your child with instant gratification makes you feel good, maybe even better than they do. But don't reward manipulation, pouting, crying, and guilt-tripping with toys and candy unless you want your child to grow up to be a reality TV star.

Talk to your kids about what really defines a person's worth. Be a good role model. Don't celebrate your triumphs with a shopping spree—bake a cake or go for a bike ride instead. Conversely, don't punish your kids by taking away their stuff—this will only teach them that their stuff is really, really important. Adjust their privileges instead.

There's nothing wrong with giving your kids nice things if you can afford them. But help them earn big-ticket items instead of taking them for granted. Give them daily chores to do, and assign a point system for each chore. Once your child reaches a certain number of points, reward her with something special, along with the satisfaction that comes with delayed gratification. Remember that buying them material goods will not build a bond with your kids—spending time together will. And since experts have filled books on the matter, I refer you to Betsy Brown Braun and Wendy Mogel, two of my favorites on the subject.

You need some time alone.

If you're afraid that telling your family to make themselves scarce will make them think you love them less, consider how they'll feel when you're crabby, snappy, and mean because you can't get a minute to yourself. Show me a human being who doesn't need to recharge her batteries once in a while and I'll show you a robot. Okay, that metaphor didn't work exactly—robots are more likely to need batteries than humans, of course—but you know what I mean. It's natural to require a bit of space.

Since you know this, plan ahead. If you schedule some time for yourself each week instead of demanding it when you feel cornered, your family is less likely to take your solo time personally. And encourage them to take time for themselves as well. Tell your husband to have a blast at the game; buy your sister a massage; teach your kids the pleasure of sitting quietly alone with a book.

Your kid is . . . how can I say this gently? . . . a dork.

Kids are supposed to think their parents are dorks, not the other way around. Stop perpetuating the *Mean Girls* social caste system—don't judge your poor little dork! So what if your kid doesn't care what sneakers he wears, or which song is number one on iTunes? If he's not wasting time obsessing over pop culture, chances are he's spending time doing something more productive. Like thinking for himself.

Still, if you're worried your kid's differences might be alienating him socially, there's no harm in extending a helping hand. First, figure out what the root of the problem is. Would your little one prefer to spend more time playing Settlers of Catan online than playing sports in real life? Then gently coax him outside once in a while. Sun is addictive, and it's too bright to see the computer screen out there. Does your kid seem to yearn to be part of the cool crowd but has no idea he doesn't look the part? Sneak a cool pair of sneakers into his closet—tell him you mistakenly ruined his other ones. Does your kid seem blissfully happy in his dorkiness? Does he radiate cluelessness and a lack of preoccupation with reality? Then you have one lucky kid. This crazy state is called innocence. Keep him in it for as long as you can.

Your kid bullies another kid.

The short answer is, if your child is a bully, it's your responsibility to find out why and to stop it. You have an obligation to teach him or her that inconsiderate behavior will not be tolerated. Setting your child on a new and better path is a process requiring time, insight, and commitment. There's no overnight fix or instant solution, but here are some guidelines to get you started.

Try to figure out why your child bullies. Ask your child why she behaves that way. Listen carefully to the explanation. You always want to be on your child's side first. While it's possible that your child is in the wrong, you're still her number one supporter, and you have to listen to what your child has to say in defense of her actions. Listening leads to understanding, and understanding can lead to change. Your child might get defensive and say that she's bullying someone because that person is annoying or she doesn't like that person, but that excuse is usually just the surface answer for the underlying reason. Try hard to get to the root of your child's anger and upset.

If your child is physically violent toward other children, try to figure out where he learned that this type of behavior was acceptable. Is there violence in your home? An older sibling who is physically violent toward him? Does your child spend a lot of time watching violent television programs or movies or playing violent video games? Does he listen to music that has violent messages and song lyrics? You have to stay on top of what's going on in your kid's life so that you can keep violent images out of it. If there's violence in your home, then that is a major problem you have to deal with as a family unit. If you were formerly in a situation of domestic violence and your child was witness to it, then you should seek professional counseling for your child so that he can learn to express anger without physical violence. Explain to your child that hitting, kicking, punching, fighting, and name-calling are intolerable behaviors that you do not support.

Make your child reflect on or her actions. Ask how she believes she would feel if someone were to beat up on her or make fun of her on a regular basis. Tell your child to put herself in the shoes of the people that she is bullying.

Demand that your child apologize to the children that he has bullied. If your child has gotten in trouble with his teacher or the school administration as a result of being a bully, he should also be required to apologize to those people.

Tell your child (and demonstrate) that if there is something she wants to talk about with you, you are available at any time. Often, a child who is a bully has low self-esteem and self-confidence. She feels the only way to gain control in peer situations is through aggression.

Help your child develop social skills by getting him involved in youth groups or sporting activities.

Nurture your child's talents and dreams so that she does not feel the need to assert herself through bullying, fighting, or making fun of others.

Now, if your child is being bullied (and let's face it—in life we are usually at one time either a bully or the bullied), then as much as you might like to walk in and crush that little brat, that will not solve the problem. Help your child figure out ways *he* can deal with the bully. Help him problem solve. Role play the situation. Obviously, if it is physical violence you will need to intervene, but if it is a situation in which your child can stand up for himself, then it will go a long way for him to learn how to do it. Let him figure out what he could do in the situation. Also explain that people usually bully because they don't feel good inside and bullying others helps them feel better. Depending on where this is happening (school sports, etc.), you should let the school or parent know, but also let them know that your child is trying to work it out on his own.

PART 5

Sticky Situations in Parties

When I was in college, an entire wall of our apartment was devoted to photographs of people losing control at our frequent soirées. We called it the Wall of Shame. From pretty girls throwing up on Laura Ashley bedspreads to geeky guys mooning to George Michael songs, the photos we posted were a chronicle of everything parents fear when they send their kids off to college.

Once, in my twenties, I was thrilled to attend a party at the loft of a very glamorous girl I knew casually but had always admired. The guest list was luminous, drawn from the worlds of fashion, beauty, film, and media. She opened the door with grace, wearing the most incredible, bias-cut silk Calvin Klein gown, and handed me a Bellini.

After politely introducing me to a couple of people, the hostess disappeared to tend to the music. I was so nervous around this group of fashionistas that I leaned into a dark corner with my drink to watch the crowd. When I tired of partying alone, I headed off to the bathroom to freshen up. As I opened the door, I had to kick a shoe (Manolo) out of the way and when I did, I realized it was attached to . . . the hostess!

There she was in all her glory, splayed unconscious on the floor, eyelids fluttering, Calvin Klein dress puddled around her, revealing her hairless privates (thereby introducing me to one more thing It girls had—or should I say didn't have—that I hadn't known about) in all their glory. I was so shocked I forgot to close the door behind me, and another partygoer—whom I like to call the Greatest American Hero—stepped in.

He quickly covered her private parts up with a towel and began to check her vital signs (luckily, he was in medical school). After she was deemed passed out and not dead, he carried her into her room and the party went on as before. I, needless to say, slipped out the back door. The glam hostess never once brought up the incident—and rather than tarnish her reputation, it, bizarrely, added to her glamour.

What is it about parties that brings out everyone's naughtiest behavior? Is it the booze? The music? The social anxiety?

Whatever the reason, so much can go wrong at get-togethers that we obviously need to devote an entire chapter to them.

Throwing Parties

You've invited a bunch of people over. Now what?

You can make a simple, largely unplanned party. Invite an assortment of people and supply the food and liquor while they supply the noise. But without too much effort or too much money, you can also create a memorable splash that is more sophisticated and imaginative. Magazines, television shows, books, and the Internet are great sources for ideas and inspiration.

Whether you decide on brunch or black tie, potluck or haute cuisine, here are a few tips to help you event plan like a pro:

- Send an electronic invitation to save trees and cash. I really like those offered on Pingg.com. I also recommend using the event features on MySpace and Facebook—they make managing your guest list really easy and give guests a way to communicate with one another.

- Be sure to let your guests know up front all the necessary information and details: formal or informal, indoors or outdoors, wine and liquor provided or BYOB, nibbles only or a full meal, kids welcome versus adults only, themes, costumes, a special occasion being celebrated, etc.

- Organize, organize, organize. Make to-do lists. Shop early for nonperishables. Set timetables for food preparation, table setting, etc.

- Don't get carried away with menu and decor. You're not launching a theme park. Your focus should be on creating a fun, enjoyable experience for your guests. Unless your name is Rachael Ray or Nigella Lawson, serve uncomplicated food that can be prepared mostly in advance, then finished off at the last minute.

- Round up when it comes to amounts and have extra bags of chips or pretzels on hand in case you run low on food.

- Have an extra cooler filled with ice for beverages. Otherwise, they'll overwhelm the fridge. Make sure to include a variety of nonalcoholic options.

- A couple of days before the party, clean the house. (Hint: clean well enough but don't go overboard—you don't have to exhaust yourself dusting every tchotchke. And don't be afraid to keep certain doors closed.) Clean the bathrooms the day of the party.

- Make sure you have enough chairs, plates, cups, napkins, glasses, and silverware for everyone.

- Don't forget music! Candles!

- Delegate. Ask key people in advance to pitch in and help greet, bartend, pass hors d'oeuvres, or be your sous chef in the kitchen.

- The day of, start setting up early and allow lots of time for everything, including getting yourself ready. No dashing upstairs to get dressed ten minutes before the doorbell rings.

- Let go of the idea that anything, let alone everything, has to be perfect. Parties should be fun. Embrace the spirit of Samantha on *Sex and the City,* who once said: "I don't believe in the Republican party or the Democratic party. I just believe in parties."

You're having your first grown-up dinner party. Um, how do you set the table again?

If you're having trouble remembering the order of the silver, perhaps it's best to skip the fish course. The general rule with utensils is to begin at the outside of your place setting, then work your way toward the dinner plate: to the right, soup spoon on the outside, then fish knife, then dinner knife next to the plate. On the left, fish fork on the outside, salad fork in the middle, then dinner fork. Set the water goblet directly above the dinner knife. Glasses for white and red wine glasses go next to it to the right, in that order. The Web site Table-settings-with-pictures.com/etiquette-for-proper-table-setting.html will show you how everything is supposed to look.

One thing to remember is that this isn't your great-grandmother's Victorian dinner party. Yes, there are "rules" about how to correctly set a table, but Modern Girls find ways to make everything their own. So feel free to play a bit and create your own personal style and aesthetic. For instance, napkins traditionally are set folded to the left of the outside fork, but there is no reason not to do your origami thing, or swirl them out from a wine goblet. Interesting napkin rings are also a great way to make your own individual statement. A friend of mine has a collection of napkin rings, from jewel-studded gold wires to faux tortoiseshell dragonflies. They make every meal at her house chic and fun. Or deliberately mix china and glassware. It's the twenty-first century. Everything doesn't have to be matchy-matchy.

Dinner Party Disasters

Stinky Fridge

Make sure there's no condensation inside, which can lead to stinky (and toxic) mold. Remove the kick plate—that metal, vented thing at the bottom of the appliance—and vacuum it. Wipe all inside surfaces of the fridge with a damp sponge coated in baking soda. If the odor persists, the problem may be in the freezer. Pour kitty litter into a box that fits in the freezer and let it absorb the freezer smell. Doing this will also make your ice cubes taste much better.

Broken Ice Maker

There's no time to waste trying to fix the stupid thing—that's what the refrigerator repairman is for (and don't hesitate to call one, especially if you're single—I have a friend who met her husband that way). What you need to do is make ice, stat.

First things first: put all unrefrigerated party drinks in the freezer. This will speed-chill them faster than putting them in the fridge.

Next, make ice manually. Refrigerated water will freeze more quickly than room temperature water, so if you keep a Brita pitcher in your fridge, pour its contents into your ice trays and refill. If you don't have ice trays, you can use shot glasses or even day-of-the-week pill containers (the kind your grandpa uses—just be sure they don't have any meds in them).

Then, mass-text your guests and tell them anyone who brings ice gets a free drink.

Refrigerator Stops Working

If your fridge turns off because of a power outage or some other reason that's (hopefully) temporary, keep the door shut if you can. This will trap cold air inside and keep food from spoiling. According to the USDA, food is safe for four hours this way. Your freezer, amazingly, will keep items frozen for up to forty-eight hours after the power goes out (if it's full; twenty-four hours if it's not).

Oven Won't Turn On

There are too many variables as to why this might be happening—depending on what kind of oven you have—to speculate about causes here. Instead, here are some quick fixes. If you have a toaster oven, chances are you use it only for toast. Use the oven part! With the exception of a giant piece of meat or a massive pizza, most things that can be cooked in a regular oven can be cooked in a toaster oven. If your toaster oven has a convection feature, set it to a temperature 25 degrees cooler than the recipe calls for, and don't let any food sit too close to either the upper or lower heating elements (at best it will burn; at worst it will start a fire). If you are cooking a whole chicken that won't fit in the toaster oven, you have two options: cut it up or throw it in a Crock-Pot or pressure cooker. You can also, of course, cook it on your stovetop if that's still functioning. And don't forget about your microwave!

You Find Out, at the Last Minute, That You Need to Feed a . . .

Vegan

Think about all of the foods that contain something that comes from an animal—not necessarily meat, but eggs, milk, cheese, lard. A vegan can't eat any of them! Sorry to rub it in.

Don't try to make a separate dish for your vegan guest. Instead, compose a plate for him or her that contains any animal-free sides you're serving, as well as something easy to prepare from your pantry, such as pasta with garlic and olive oil tossed with lemon zest, rice and beans, or couscous with almonds and raisins. There is no need to make a last-minute mad dash for Tofurkey. I like to compose vegan guests' plate, rather than have them serve themselves from family-style dishes, so that they get enough food (and so that they don't hog too much of sides that are also supposed to feed the meat-eaters).

Vegetarian

See above, except eggs and dairy are A-OK.

Lactose Intolerant

The only thing off-limits for this poor soul is milk. Other dairy, such as yogurt or cheese, is probably okay in small amounts.

Sufferer of Celiac Disease

Those with celiac disease need to follow a gluten-free diet. Gluten is found in wheat, barley, rye, and sometimes oats, so don't feed your guest any of these. A simple protein and veggie dish won't bother their system.

Diabetic

Diabetics are sensitive to foods that raise their blood sugar, which means they usually avoid simple carbohydrates (like sweets or white bread) and need a strong foundation of protein and complex carbohydrates (whole-grain pasta is an example). Most diabetics can eat fruit safely, so it's a good option to offer for dessert—even for those who don't have the condition.

Food Allergy Sufferer

This is one time you don't want to fib about ingredients. People with severe food allergies—typically to nuts, seafood, legumes, milk, or eggs—will go into total systemic shock, called anaphylaxis, if they accidentally consume even a microscopic particle of the food they're allergic to. If this should happen, find out if they carry an EpiPen, and call 911 immediately, as they could stop breathing within seconds. Never serve anyone with food allergies any dish whose components can't be clearly identified—a soup, stew, or sauce could contain bits of shellfish, for instance, or a pesto might have ground nuts as an ingredient. Even if you don't suspect you have an allergic guest at the table, present your dishes like you're on *Top Chef*, explaining their preparation and ingredients, and you'll avoid unwanted trips to the emergency room.

Sticky Note: When inviting guests for dinner, try to remember to ask about any dietary restrictions at the time of the phone call or e-mail. This way, you can avoid having to throw together a substitute dish at the last minute.

Sticky Note: Keep on hand a stash of vegan entrees from Trader Joe's or an ethnic grocery store. When you have unexpected vegetarian guests, prepare according to the package directions and add some fresh vegetables and herbs—whatever you're putting in the meaty main course—and perhaps some tofu if you have it. They will be touched by your consideration.

An impromptu band of guests has shown up, and you have nothing to feed them.

I was once snowed in at a friend's country house and totally depressed at the thought of having to eat oatmeal and canned soup for days on end. Boy, did she surprise me! The magical dishes she made out of cupboard staples and neglected fridge basics tasted fresh and totally gourmet. Here are some of her genius concoctions.

Pasta with anchovies, olive oil, and crushed red pepper. Sounds gross to some, I know, but heat some olive oil, add anchovies and mash till they dissolve, then toss with hot pasta and sprinkle on crushed red pepper. You'll think you're in a trattoria in Napoli.

Salmon cakes. They can be as delicious as the kind you get in a fancy restaurant, I kid you not. Just drain a 15.5-ounce can of salmon, then mix the salmon with an egg, 2 teaspoons mustard (I like Dijon for adults and honey mustard for kids), 1¾ cups bread-crumbs, and one stalk of celery if you have it. (The celery can be that sad, limp, has-been-living-in-the-fridge-for-weeks kind.) Form the mixture into eight generous patties. Cook in canola oil over medium heat for about three minutes per side, or until golden brown. Serve with a simple green salad if you have it. If you don't, make twice as many and present these little guys as an hors d'oeuvre.

Corn muffins with fresh corn. Just prepare a box of corn muffin mix—old-school Jiffy is my favorite—and add a cup of frozen sweet corn to the mixture before baking. If you want to get fancy, add 3 tablespoons of chopped basil—dried is okay, but fresh is better. If fresh, use 1 tablespoon less.

Polenta pizza. If you have cornmeal, make it into a mush, then spread along a greased jelly-roll pan like a crust. Place in the refrigerator and allow it to cool and harden. Then, remove it from the fridge and cover it with a thin layer of jarred tomato sauce, black olives, and Parmesan cheese. You can sprinkle a few shakes of crushed red pepper on top if you like it spicy. Bake in a 350°F oven until golden brown. Serve cut into hearty rectangles, like lasagna. This is a great option for people who don't eat wheat or are on gluten-free diets.

Sticky Note: One of the best Web sites I've found in a long time is Supercook.com. It allows you to input a list of ingredients you have, and then spits out a list of recipes you can make using those ingredients. For example, typing in "zucchini, chicken, cloves, wine" yields Caribbean Chicken and Vegetable Kabobs. Yum!

You're trying desperately to get the dinner ready, and your guests won't clear out to let you work.

First, try distracting them by leading them into another room and providing an unexpected activity there. Like, "Have you guys seen my Ms. Pac-Man machine? Go ahead, try it!" If that fails, enlist a trusted guest (your beloved, your BFF) to entertain the others while you focus. If no one wants to give you your space, ask them to help you with a project that can be done in another room, like setting the table or taking out the trash. Come to think of it, asking guests to take out the trash will surely clear out the kitchen.

A guest, trying to be helpful, breaks a piece of your grandmother's china.

As tempting as it is to shame that person by whimpering, "That was the only thing salvaged from the house of the grandmother I never had a chance to meet," you must be gracious and make the hapless guest feel as unguilty as possible. Grab a broom and dustpan and get rid of the mess as soon as possible. If your guest is truly distraught, break something yourself so you're even. And be sure to save the broken pieces—you can use them to make a mosaic, or to guide you in finding a replacement online at Replacements.com.

A guest insists on taking command of a dish and subsequently ruins it.

Never let a guest take charge of your kitchen unless you know he or she is a better cook than you are. If it's too late, just be sure everyone knows you're not the one who made the flat soufflé—drop backhanded compliments such as "I'd like to toast Susan, who really helped me today by taking the lead on the soufflé."

Cocktail Hour

You're no expert, but you're pretty sure the wine is "off."

Essentially, there are four things that will make a bottle of wine taste off: being corked, oxidization, maderization, or refermentation. While the aforementioned terms aren't something you'll hear every day unless you're a sommelier, if the red wine tastes carbonated or if the first sip of wine is reminiscent of mildew or vinegar, chances are you need to open another bottle. Save this one, take it back to the wine store, and they'll give you a credit.

Your guest asks for a sidecar, a Manhattan, or a Tom Collins (or—cringe—a Sex on the Beach), and you have no idea how to make it.

Ask your guest to teach you how to make one. If you're too shy, jump online and check out Webtender.com, which has a massive database of cocktail recipes. Or if you have an iPhone, download an app such as iBartender, which will tell you how to make whatever your guests fancy.

First Course

You burn the bread while reheating it.

Using a sharp knife, scratch off the burned surface, and you will have perfectly heated bread minus the unsightly burn.

Your bread is stale.

Unless your bread is growing blue and green mold, there is no reason why you can't eat bread that is a few days past its prime. The best thing to do with stale nonmoldy bread is to turn it into something crunchy and delicious, like crostini or croutons for salads. You can also refresh stale bread by sprinkling it with water, wrapping it in aluminum foil, and heating it in the oven—a chef's trick that makes dinner guests think it came out fresh from the oven.

The Main Event

You burn the meat.

If you've scorched only the surface, wrap the meat in a towel for five minutes to steam the burn off, then scrape off the char with a butter knife. Slice the meat and arrange it on a platter to minimize any imperfections.

If the whole thing is dried out, shred the meat and mix it with salsa, sloppy Joe sauce, or barbeque sauce, and serve it as a filling for sandwiches or tortillas. Or dump it, along with some finely chopped sautéed carrots, celery, and onions (fancy folks call it mirepoix), into a pot of store-bought stock for a hearty soup. Serve with crusty bread, or toss in some barley or rice for a full meal.

Although sometimes easier said than done, a good rule of thumb is to keep chunks of meat large so the exterior can brown before the interior dries out. If the damage has already been done and you have mouths to feed, conceal the problem with an olive-oil-heavy sauce (to help give the illusion of moisture) with chopped fresh herbs. Another alternative for really dry meat is to chop it into a salad or shred it into tacos.

The chicken is raw.

The good thing about raw chicken is that it's easy to identify. Poultry is undercooked when its juices run bloody or cloudy—they should be clear. Unlike red meat and fish, poultry should not be eaten raw because it carries a high risk of salmonella.

Before taking a chicken dish out of the oven, pierce the meat at its thickest part to release the juices and check for doneness. If you don't trust your ability to judge the color of your chicken's juice, stick a meat thermometer into the meatiest part of the flesh—it should read 165°F. Novice cooks may want to stick to recipes that call for chicken pieces instead of whole birds, since they cook more evenly.

So what if you discover the chicken's undercooked once it's already on the table? Don't attempt a cover-up—you and your guests could get really sick. Instead, crack a joke—"No wonder I didn't get that job at KFC," or "I don't know about you, but I'm too chicken to eat raw chicken."

Obviously you can't throw everyone's partially cooked meal back into the pan, so you're going to have to get resourceful with a quick and easy replacement course. Pasta is a no-brainer, but what if you don't have any sauce? All you need is some garlic, olive oil, and crushed red pepper. To flavor a pound of pasta, slice a couple of cloves of garlic superthin, then sauté them in ½ cup olive oil until they just barely start to brown. Add ¼ teaspoon of crushed red pepper, then pour over the pasta and toss. Extra credit if you have some finely chopped parsley to add for color.

If you don't want to do pasta, make eggs. You can go fancyish, with mushroom omelets, or comforting, with scrambled eggs and cheese. Garnish with fresh herbs and serve with a simple salad and whatever side dishes you planned to serve with the chicken. Eggs make a satisfying meal that feels surprisingly sophisticated in an effortless, European way.

If there's absolutely nothing left in your kitchen, break out a new bottle of wine and order a pizza. Or dash out to the grocery store and snag whatever's spinning on the rotisserie. Next time, skip all the drama and make this last resort your step one.

Everyone wants dinner—but you forgot to defrost it.

The best way to defrost is to place the item in the fridge and let it naturally thaw out for two days in advance. However, if you are in a time crunch and need dinner ASAP, cold water works well, too. Simply run cold water over your frozen soon-to-be-dinner and voilà—it will be ready to cook a lot sooner.

Food Safety

The expiration date is illegible. How do I know if it's spoiled?

Everything spoils at a different pace, and foods have much greater longevity when stored in the freezer versus the refrigerator. If you can't read the expiration date on refrigerated meat or dairy, it's best to say goodbye and move on. When trying to determine whether veggies are still good, consider factors such as color and smell. When in doubt, throw it out.

You forgot to refrigerate your leftovers, and now you can't figure out whether it's safe to eat them.

Do yourself a favor and put the fork down. When food products are left at room temperature for more than two hours, they are at risk for harboring potentially harmful bacteria.

You oversalted—like, enough to fill an ocean.

The best scenario is if you oversalt before you begin to cook. In this case, all you have to do is wash off the salt and you can begin again with a blank canvas. If it is after the fact and you are about to serve dinner and realize you have oversalted, adding more liquid or solid ingredients will help dilute the saltiness. If things are really bad, an old chef's trick is to counteract the saltiness by adding a few pinches of sugar or a teaspoon of vinegar.

Your food is so bland it could be served in a hospital.

In the kitchen, this is the easiest problem to solve as long as you don't let yourself go too crazy. Yes, foods without flavor are not as satisfying, but when it comes to seasonings the simplest dishes are often the best, and unless you're a professional chef, stay on the conservative side when you are adding spices.

Your sauce is runny.

To add substance to most runny gravies or cooking sauces, place the liquid in a pan of hot water and once simmering, add a paste of cornstarch and water to thicken.

Just Dessert

Your cake collapsed.

The most likely reason a cake collapses is that it has not cooked through. Sticking it back in the oven, sadly, will not work, as baking is a precise science. You best bet is to make a new one if you have the time and the ingredients. If not, break off the cooked parts and make a trifle with them—layer crumbled cake, fruit, and whipped cream in a big bowl with a flat bottom. Add a little booze—rum or Bailey's works well—and nobody will care what you were trying to make in the first place.

If the ill-fated cake was intended for kids, not boozers, press its crumbs into the bottom of a pan and top with a smooth layer of softened ice cream. Top with nuts or chocolate chips and freeze. When you're ready to serve it, warm the sides of the pan to release the cake, and you've got an ice cream cake that would make Baskin-Robbins proud.

After Dinner

None of your eight thousand guests is offering to clear the table.

I have a friend who throws tons of dinner parties, and at the end of the meal, when she's frantically trying to move all the dirty dishes into the kitchen so she can bring out the dessert, her guests are usually lazing about the table, swilling their wine and arguing over rock bands or reality shows. I'm always amazed she doesn't throw her hands up and say, "Hello? People? Can someone carry in a dish?" But she stays composed. I think she enjoys being totally in control and serving people.

I, however, don't. I welcome any help. And I quietly seethe with resentment when my guests don't offer it. Instead of being passive-aggressive, I find the best strategy for building my kitchen-cleaning army is to declare, with a wink, that whoever brings the most dishes in gets the biggest piece of cake.

Sticky Note: Because flavor and fragrance are so closely linked, never burn scented candles or wear strong perfume when you're cooking or serving food. Doing so will throw off the taste of the food you've worked so hard to make.

Dear Jane,

What makes something taste good when you can't necessarily identify the goodness? While I know what I like, I have absolutely no innate sense of how to cook. Is there any hope of me developing a culinary instinct if I've never had one before?

Many of the best chefs I know say that the key to cooking well is not a French culinary education but a good palate. Don't be afraid to experiment with recipes as long as you're tasting all along the way.

Chefs identify foods as having one of five types of flavor: sweet, sour, bitter, salty, and savory (also known by a Japanese term, *umami*). The art of cooking well lies largely in combining these flavors. While there's always going to be a certain amount of trial and error, I swear by a book called *The Flavor Bible*, which offers an index of ingredients and pairings. It might surprise you, but bacon and chocolate are BFFs.

You have an uncontrollably drunk guest.

Every soirée has one . . . the guest you could *so* do without! People who fall into this category are either loud and obnoxious, rambling on about how they're your best friend, picking fights with other guests, or threatening to be sick in someone's lap. Quickly, quickly, before someone like this blights another minute:

- Get the person away from any more alcohol. He will probably want to continue drinking, but your job is to run interference. Get him drinking as much water as you can. It's hard to convince a drunk to do anything he doesn't want to do, but the more water he drinks, the better he'll ultimately feel. Water will at least keep him hydrated, as being dehydrated adds insult to injury.

- Offer high-protein or high-carb food.

- Steer her clear of exes and anyone she has reason to argue with. However much she dislikes a person while she's sober, her emotions will be multiplied with alcohol in her system.

- Get a drunk guest home, but *never* allow him or her to get behind a wheel. Have a sober guest drive, or call a taxi and pay for the ride yourself.

When the Drunk Tanks

FYI, the social downside of a drunk guest spoiling a party is only half the problem. In many states, you can also be held responsible for what happens to your drunk guest after he or she leaves your home. It doesn't matter if he lands in the hospital after falling down a flight of stairs or if she drives away drunk and causes a car crash. If you allow your friend to get drunk, you could find yourself in court. And even if your friend doesn't sue you for damages, a person he or she injures in a car crash could. According to the Insurance Information Institute, thirty-seven states have adopted some form of "social host" law or set a legal precedent that allows you to be found liable if a guest injures himself or someone else as a result of alcohol consumption on your premises. Some social-host laws have conditions. For example, in South Carolina and Nevada, liability applies only if your guests are under twenty-one years old.

But in any and all cases, it is important to pay attention to how much your guests drink.

Six people are showing up in ten minutes, and you have no time to go to the store.

You can't pull out those truffles you brought back from Italy last week? Fresh, shrink-wrapped cheeses from that famous Paris fromagerie? You're not the sort who keeps champagne chilled and at the ready, just to be prepared for an emergency like this?

Three words: *refrigerator, cupboard, garden*. Four more words: *fruit, vegetable, drinks, bread*. All under the headline of improvisation. Work with what you have on hand.

- Section oranges or apples and put them on a glass plate decorated with sprigs of dill or parsley. Include cheese if you have it on hand.
- Put berries in a bowl. If they're strawberries, include a second small bowl of sugar for dipping.
- Make a pitcher of iced tea or iced coffee. And of course pull out any wine and sparkling water you have. Use mint or rosemary as garnish.

Later, in the aftermath, decide whether it's a one-time, forgivable lapse or if you're erasing this person from your contact list.

Your best friend asks you to throw her baby shower or bridal shower, but you're broke and clueless.

Two golden rules: (1) Thou shalt rise to the occasion when a BFF has a life-changing event that must be celebrated. (2) Thou shalt never be a hostess of cheap or cheesy fêtes. Amen to both.

Actually, there's a third golden rule at work here: (3) Thou shalt not reinvent the wheel except when absolutely necessary. In other words, people with far greater experience than you in these matters have blogged and published on the subject, and you should take advantage of their advice and expertise.

Over my many years of throwing and attending showers, I've picked up some handy money-saving tips:

- You don't have to serve a meal. Have the party at an odd hour to send a clear don't-expect-to-eat-much signal. Even better, throw an ice-cream social. Cheap, and who doesn't want an excuse to gorge on butter pecan?

- Specify an ending time on the invite—i.e., instead of writing "2:00 P.M.," write "2:00 to 4:00." The shorter the party, the less it will cost.

- Ask each guest to bring a balloon. Voilà! Instant decor.

And don't feel embarrassed about pinching your pennies. In this case, it really is the thought that counts.

You forget to invite someone important to your party, and he or she finds out.

If the party hasn't happened yet, call right away and invite the person. Be honest—that you intended to ask her but forgot. Offer an excuse, lame or not, about the craziness of your life lately and the other people and things you've been forgetting. Apologize so profusely she will be convinced you are sincere (which, hopefully, you are). Within a few days send a note, card, or e-mail offering another brief but graceful mea culpa for your oversight. If she's coming to the event, say you're looking forward to seeing her, and reiterate the particulars of time, place, and so on. If she can't attend, say you're looking forward to making a date to get together soon, and then follow up on that.

- Arrange small bowls of nuts, seeds, and olives on a tray.

- Julienne carrots or celery. Break up cauliflower or broccoli into small florets. Make a quick dip by adding minced garlic, lemon juice, minced chives, dill, or all four to mayonnaise. Or put out sea salt.

- Commandeer available breads and spreads and cheeses, from Carr's Table Water Crackers to Wonder bread. If worse comes to worst, you can put butter and Parmesan on squares of stale bread and pop them under the broiler until they're brown and bubbly.

- Scour the cupboard for cookies, chocolates, even leftover jelly beans from Easter.

The point is, Modern Girls may not be domestic goddesses, but we are not without the resources or smarts to pull a rabbit—preferably chocolate—out of a hat on a moment's notice.

If she finds out after the fact, approach her and tell her about your oversight. Express your embarrassment and invite her to dinner or for a cocktail so she knows you do actually want to spend time with her.

You never know what music to play.

Ask another guest with great taste to bring his CDs, iPod, or MP3 player. Or, if you're feeling democratic, ask each of your guests to bring a favorite CD along, and play them in the order of each person's arrival. It's fun to guess whose song is whose.

You've invited an interesting group of people, but you have no idea how to seat them.

Two cardinal rules are (1) divide couples up and (2) don't seat people next to each other who you know are oil and water (the gun control activist next to the National Rifle Association member who thinks pistols should be given to each American child). Conversely, definitely seat together people who don't know each another but who you know to have interests in common. Or make it man-woman-man-woman around the table, to keep the flirtation factor high. Nigella Lawson, British author, TV personality, and self-proclaimed domestic goddess, always begins with those sitting nearest to her. "Sometimes that will be the most difficult person to place," she says. "But sometimes it will be the guest of honor." Generally speaking, it makes sense to start with the people who concern you. You can take your time with the charmer—he or she can sit anywhere.

One of your guests is trying to clandestinely gather up your troops and leave your party for another one.

Whatever you do, don't beg them to stay. It may seem dumb to play games, but you'll be miserable if you think everyone is hanging out at your place just because of guilt. Just be cool and say, "Have a fantastic time! I really enjoyed having you, and I know they will, too." If you're friends with the person whose house your guests are headed to, you can always invite their party to crash yours (provided your place is big enough and you have enough refreshments). You can do something to make them want to stay, like break out a massive platter of ice cream sandwiches, or start a rollicking game of Celebrity. Or you can reward the guests who do stay with a really special glass of dessert wine and a concert featuring your collection of old soul records. If worse comes to worst and your soirée ends earlier than you'd anticipated, consider yourself lucky. You can clean up before you go to bed and start the new day like nothing ever happened.

A friend promises to show up early to help you get ready, then bails.

First you need to deal with the party, then the friendship. If you have enough tasks for four hands but can only avail yourself of two, prioritize. First and foremost, clean. Throw clutter into a closet and shut the door—you can deal with it later. Wipe down all kitchen and bathroom surfaces with fresh-smelling spray cleaner (I like the Method ones from Target the best—they're all natural and smell that way). Place fresh flowers wherever possible—they create the illusion of spotlessness. Make sure you have ice, and that drinks that are supposed to be cold are in the fridge. If you have anything that needs cooking, do as much prep work as you can beforehand. Then put on some soft music, pour yourself a glass of something indulgent, and get ready. Put your hair up, since you're going to be running around and you don't want to be flushed and sweaty when guests arrive. For that reason, skip the foundation and go with simple mascara and lip gloss (you can always add powder later, once you're settled). Wear something cool and easy that doesn't require uncomfortable shoes. Put an apron on immediately so you don't get muck on your frock. Once you're presentable, you can return to the minutia of party making, like lighting candles, unwrapping cheese, and opening wine.

When the friend who bails arrives to revel, give her a chance to apologize. She might have a perfectly acceptable reason for failing you. If, however, she acts as though nothing is wrong, call her on her mistake and tell her how frustrated and disappointed it made you feel. She may feel awful for a minute, but don't you want her to, just a little bit? Then the tension will evaporate and both of you can enjoy the party. And chances are she'll stay to help you clean up.

You want to move the party outside, but there are mosquitoes everywhere. And you have no repellent.

Give each guest half a sheet of fabric softener to put through their belt loop or tuck in their bra. Mosquitoes hate the smell of it. (You've never found them in your dryer, have you?)

You want to ask a guy you have a crush on to come to your party, but it's weird.

Okay, first, answer this question honestly: are you throwing this party just so you can hang out with this guy at your place? If the answer is yes, save yourself a couple hundred dollars and ask him out on a date. Sure, you may want him to experience the awesomeness of your friends, apartment, and book collection, but none of these things matters if you two lack a connection.

If you simply think this dude would make a delightful addition to a party you're already having, then by all means invite him. Calling is the most elegant way to do it—texting seems shady, and e-mail may not be personal enough to coax him to the house of someone he barely knows. When you call, give him some low-pressure reasons to attend: you have this friend in common, you want to show him this awesome flea market find, you made way too much of your famous chili and are desperately searching for people to eat it. Even if he declines your invitation, you've opened the door to more possibilities.

Someone you didn't invite sees your Evite list.

How could she not, with you checking it every five minutes from your work computer? If you don't want your party to be public knowledge, you may need to go to considerable effort to keep things under wraps. Don't talk about it at work unless you've invited everyone in the company. Be sure to set the invitation and guest list to private if you're using an online service. Don't shop for a party dress at lunchtime. And don't solicit advice on planning decisions. If the uninvited person creates an uncomfortable moment of confrontation—maybe she says something like "Gee, that bouncy house you're renting this weekend sure sounds like fun" or "I didn't receive your invitation, so can you resend?"—it's best to just invite her unless she's going to ruin the shindig. If the fiesta is particularly teensy or particularly fancy and you really can't squeeze in one more person, you shouldn't feel the least bit guilty telling her that while you had to make some tough decisions for this party, you can't wait to welcome her to your next rager.

Parties: Going to Them

You want to see if a friend was invited to a party you're going to.

Hmmm. Since the first rule of friendship, like the first rule of medicine, is "Do no harm," this could be a tricky maneuver unless you're in a Jennifer Aniston movie where everything turns out swimmingly in the end. Otherwise, it's real life, and embarrassment and hurt feelings can linger. You can drop hints to your friend, like asking if she's made an appointment for a mani-pedi a week from Saturday. Or you could ask the host if there are other guests coming from your part of town, as you were thinking about carpooling. If you drop clues like crumbs along a path and no one follows them, don't pursue it. We never quite outgrow the pain of being excluded, and the last thing you want is for your friend to feel left out and you to be the source of it. Should she not be invited and asks if you are, tell the truth. Lies, even white ones, are a form of doing harm, too. You may rationalize that you're saving her from discomfort, but actually, it's yourself you're saving. Better to keep it clean and honest and soon after make a special day with your friend.

You have to go to a party where you'll know virtually no one.

You can definitely pull this off. Going to a party by yourself can be daunting, but here are a few tricks that will help put you in the right mind-set to meet many new people and have a great time.

First, wear something you feel fabulous in but which is also comfortable. Constantly fidgeting or adjusting your clothes will make you appear nervous and distract you from conversations. If you can, wear shoes you've worn before.

Don't arrive too early. If the party's casual, it's acceptable to show up half an hour to an hour after the start time. If the other guests will have already arrived, you'll have a range of opportunities to start conversations. But don't be too late or it'll be difficult to catch up. If the party is more formal—a dinner, for instance, where you have to be punctual—arrive just before dinner is served. Skip the cocktail hour—people tend to be settling in then and talk to people they know. Freer mingling is done after dinner.

Have topics of conversation in mind to talk about (but not to the extent that you'll sound rehearsed). Read the paper a couple of days before the party so you'll know what's in the news. Think about what your opinions are on current movies, TV shows, and current events. What books are you reading? Get a subscription to *The Week*, a veritable cheat sheet of what's going on.

Go wanting, and expecting, to have a good time. This will keep control in your own hands. Examine all the variables that make you afraid to do things on your own, and one by one take them out of your thinking. Allow yourself to be who you are, even if that includes moments of shyness, awkwardness, or butterflies in your stomach. They may be part of you, but only a part. There's also the confident person who takes chances, knowing that good people are like sunny days—there are lots of them. Who knows, there might even be advantages to going to a party alone, like not having to worry if the person you brought is having a good time.

You misinterpreted the dress code.

You can turn around and leave, but don't—that would be giving up.

You can pretend nothing is amiss and mingle as you normally would, but this takes tremendous chutzpah, not to mention it's a bit like the emperor's new clothes. All it takes is one person pointing out your gaffe and the illusion is shattered.

You can joke about it, if you're comfortable with humor where you're the punch line. Or you can turn it into a story, an anecdote that neither apologizes nor explains in stultifying detail but rather glides over the surface like a beautiful skater across a pristine icy lake. It was such a bad day at the office you forgot. Of course you know it's not Halloween, but the hosts specifically asked you

Deciphering Goofy Dress Codes

There is no shame in asking your hosts to decipher the dress code on their invite if you don't know what they're talking about. That said, here's a quick guide to the most common ones.

Creative black tie. Think Sean Penn at the Academy Awards. Your guy can wear a tux with a dark shirt and no tie. You can wear a long or short dress or evening separates. Maybe you even wear a long evening skirt with a simple but immaculate T-shirt or turtleneck, à la Sharon Stone at the Academy Awards some years back.

Resort dressy. Anything silk or linen and colorful. No loud and scary aloha shirts, but a restrained aloha or a silk camp shirt would do. Lightweight cashmere. Whimsical prints. Think Ralph Lauren.

Beach formal. Light colors, natural fibers, hat, comfortable shoes that can walk through sand (beautiful flats, or wear heels and at the edge of the beach take them off and carry them).

to provide the scary element. Invent and devise. Turn yourself into the walking definition of savoir faire.

The point is, say or do whatever allows you to be comfortable in your skin, whatever skin it is you happen to be wearing.

You don't know what to bring.

Champagne is always festive, but unless you spring for a bottle over $30, you risk giving everyone a headache. If you don't want to spring for Cristal Brut or Veuve Clicquot, a bottle of prosecco spumante, the Italian sparkling wine, would be nice, or, in the middle of a long hot summer, a good bottle of French rosé from Provence. Rosé has become seasonally chic again, especially for brunches in the dog days of August.

Sticky Note: Resist bringing flowers. By the time you knock on the door, the hostess has long since placed all the flowers for the party. If you show up with a bouquet of daisies, she'll feel obligated to stop whatever she's doing and scramble to find a vase. Sending a floral arrangement either the day before or the day after the party, however, is a super idea. (That's for you, florists of America!)

Safari chic. Nineteen-seventies YSL . . . or tons of khaki. Animal prints optional.

International. It's a small, small (business) world: suits for men, suits or business-y dresses for women.

Festive. Usually seen around the holidays, to complement the mood of a party—sparkle, beaded sweaters with black pants, pencil skirts with little red angora sweaters. Sequins may be involved, but please, no hand-knit reindeer grandma sweaters.

When in doubt, wear a black dress and keep a "costume bag" full of wacky accessories in your car.

You don't know anyone at a party.

So a neighbor has invited you to a party, but you barely know him or her? You have nothing to do and would love to get out of the house and away from your needy cats, but. . .

We've all attended events where we barely knew anyone, and yes, they can be excruciating. There is nothing quite so uncomfortable as pausing—or is that posing?—on a threshold to check out a sea of unfamiliar faces. You wonder if your time there is going to seem like an eternity; you have an irresistible urge to grab your cell phone and call someone, anyone, you know.

I'm here to tell you to get out of the house and get to that neighbor's party, and in the process go from being tied up in knots to having a blast. How to be social and get your groove on when you don't know anyone?

Initiate conversations. Watch for a group of three or four who aren't standing in a tight circle. At a party, people expect strangers to join conversations. Enter the circle by standing at the outer edge for a moment till someone notices you and draws you in. After which, you make a comment on the weather, the great canapés they're serving, or that slasher film now showing at the multiplex. Or ask questions that allow the others to open up. "How do you know the hostess?" is an old standby, but it works. A compliment can also do the trick: "Your necklace [purse, sweater, etc.] is so pretty. I've never seen anything like it." Flattery goes a long way, but be sincere. Hint: Listeners are always more in demand than talkers. After your opening gambits, perfect the art of listening.

Try to remember people's names. A good way to remember it is to repeat a name in your mind many times, and repeat the name when you talk to the person. But don't forgo talking to someone again if you forgot his name—just ask for it again with something like "I'm sorry, I forgot your name. What is it again?" Chances are the person won't be upset with you, and may have forgotten your name as well.

Keep a drink in hand—any kind of beverage. The point is to keep you from worrying about what to do with your hands. This is especially helpful if you find yourself in a lull, standing alone. If you're at a dinner party, eat slowly so that you'll always have something to do if a conversation lags. And never put too much in your mouth at once.

Smile and keep your body language open. People don't want to get to know the person leaning against a wall, arms folded, mouth downturned. The same is true of the guy grinning from ear to ear for no apparent reason. A slight, corner-of-the-mouth smile gives you an approachable look but lets you keep your edge.

You run into your ex.

Irrespective of who broke it off, running into an ex can be problematic for many reasons: lingering affection, erotic memories, the pain of being betrayed and/or dumped, unresolved emotions. When confronting your romantic past, the best protection is fabulous male eye candy on your arm, especially of the dark-haired, heavy-lidded French, Italian, or South American variety. If that's not the case, here are some tips to keep your charm and mojo high-profile.

- Don't avert your gaze. Look your ex straight in the eye and smile. Shying away from eye contact diminishes your power. Meeting someone's gaze keeps you in control.
- Be polite. Please, no histrionics—this is not *Desperate Housewives* or the third act of *La Traviata*. Fall apart later. Bitch to your girlfriends later. If you're introduced, smile and extend your hand. It may be creepy, but do it anyway. If you're not introduced, you can decide whether to present yourself or not, but in any case be courteous. While gritting your teeth, pat yourself on the back for your superior social skills.
- Take charge of the conversation, keeping the dialogue light and superficial. Talk about your dog or something that happened earlier in the day. Do not ask if he is seeing someone. Be upbeat. Enthusiasm is as much an element of control as direct eye contact.
- Don't drink too much. Emotions held in check while you're sober can spill over into angry scenes when you're chugging Jack Daniel's. You want to stay on the high road, not end up screaming at him or sobbing in the bathroom.
- Trust in your future. Tell yourself that your heart will heal and you'll be fine. Which you will.

You want to leave a party early without seeming rude.

There are ways to leave early without offending your host or ruining the spirit of the party. Number one, advance warning. If you know you're going to have to leave early, say so when you're first invited. Otherwise, explain the situation to the host—with *beaucoup* apologies—when you first arrive, or as soon as possible. Jumping up from a seated dinner party and exclaiming, "Sorry, it's a school night" really isn't the way to handle it. Number two, when you do leave, make it swift and silent. Say goodbye to the people you are talking with, then find your host. As you thank him or her, find some aspect of the party you enjoyed and point that out. The next day, phone the host. It's always polite to do this anyway, but when you've left early, the second thank-you will cover your butt.

If the urge to make a speedy exit just comes over you midparty, for whatever reason, how and when you do it depends on the size of the gathering. If it's a large gathering, you still should inform your host ASAP, giving a good reason like a work appointment at dawn, and then slip quietly away. If it's a sit-down dinner, sorry—you're stuck through dessert and the conversation *après* . . . at least a half hour. Premature departure isn't just rude, it's also insulting to other guests. Of course you can always pray the house catches on fire.

If desperate, you can always get a fake "emergency" call on your iPhone. Just download the fake call app and program for when you need. It will call and sound like a real person!

Social Damage Control

You witness people behaving inappropriately (sex, drugs, cheating . . .) in a social situation.

What you do depends on whether you know the offending parties, but by all means do something. Asserting ourselves in situations that are illegal, rub us the wrong way, or are inconsistent with our values is a way of taking responsibility—for ourselves and for our beliefs. If you're not acquainted with the parties involved, tell the host or hostess or whoever is in charge so that person can do something about it. If you do know them, confront them. No need for a lengthy lecture. You're not the police, you're a fellow citizen. Just say "hey" and remind them they're in public, which means that their behavior, though they may think it's private, is being shared with everyone.

You spill red wine on the couch, break the host's antique lamp, clog the toilet . . . how to come clean and how to fix it.

In this case, haste will not make waste; it will make the initial step toward neutralizing the problem. Tell the host or hostess immediately what you've done. Apologize profusely and offer to help in whatever cleanup or repair is necessary. Assure him or her you'll pay for any wreckage or ruin, and make good on that. In other words, fix what you break: it's the first law of damage control.

You wake up to realize you danced on the table, drunk-dialed, confessed a crush, or any combination thereof.

Otherwise known as the black hole of free expression, or what you can do only blitzed because you'd have better sense when sober. Ever wonder why shoot-outs in the old Westerns always happened next to the saloon? Alcohol makes you do really dumb things.

If you have amends to make, make them. Pole dancing at your best friend's wedding reception would fall into this category. For the more private lapses, such as eating your way through your entire pantry, just chalk it up to a bad night and move on; trying to repair the damage will probably make it worse. Next time, in addition to a designated driver, consider getting yourself a designated dialer to stop you from pushing buttons on your phone. If you simply cannot bear the humiliation, you can try to get by with the excuse of "Wow, I was so drunk, I don't really even remember what happened last night." But chances are few will believe that, and it may exacerbate the situation by allowing others to spin a tale worse than the event really was, leaving you unable to correct matters!

You wake up next to someone whose name you don't remember.

All one-night stands should come with name tags. According to a *Playboy* poll, 51 percent of men and 46 percent of women admit to having had sex within six hours of meeting someone, so chances are there have been plenty of lovers with short-term memory lapse.

If you're at his place and you don't care about him or his name, simply sneak out the door before he wakes up. If it's too late for that, decline his offer to stay for coffee and plead an early morning appointment. Do thank him for a lovely evening on your way out.

If you do like him, use some acting skills thrown in with a little private investigator ingenuity. Lying in bed next to him, mention teasingly that it all happened so quickly you feel it's now time to become properly acquainted. Introduce yourself, extending your hand as if to shake his as you sexily murmur your name. If he doesn't reciprocate, go for a hygienic approach. Tell him you like yourself and your partner to start the day with clean teeth and fresh breath. Hop up, go into the bathroom, and brush your teeth with your fingers. Hopefully, when you return, he'll follow suit and you can spring into action, checking his wallet for ID before he returns. Or when you're in the bathroom, check his medicine cabinet for prescriptions bearing his name. (Just be sure he doesn't have roommates.) If that fails, tell him you want his phone number and would he please program it into your cell? Hand him the phone and then lean over his shoulder and watch. God willing, it'll be more than his initials.

You have to go to work, and you're still drunk. Or, worse, your boss asks you why you smell like a bar and then sends you home to shower.

Probably one of the strangest, not to mention irrational and destructive, ideas you can have while drunk is that going to work will be just fine. Bottom line #1: Showing up at the office smashed and smelling like a distillery will never play out well. If you're lucky, a compassionate coworker will intercept you, call a car service, and send you home. If you're unlucky, you'll run into your boss and your job will be on the line.

That said, people seem to be falling off the wagon all over the world despite a zero tolerance attitude in the workplace. A 2007 Australian survey found that more than one in ten people admitted having gone to work under the influence, and one in three went to work hung over. And those figures were deemed low because even in an anonymous survey people worry about getting caught. A recent United Kingdom survey found almost a third of employees had been to work with a hangover and 15 percent had been drunk at work.

In the past, alcohol and the workplace often

seemed to go hand in hand, especially in creative fields. But that's no longer an acceptable case to make. In every survey, people who go to work feeling worse for wear admit to making mistakes, doing as little as possible, going home early, and struggling to concentrate. And their bosses overwhelmingly name alcohol as the number one threat to employee well-being and productivity.

Bottom line #2: Should your inebriated state cause you to see every red flag through a hazy blur and you show up at work anyway and get busted, don't go on the defensive or become belligerent. Sincere contrition and a promise (carried through on) to deal with your problem are the only way out. That, and reminding the boss about a million times of your terrific work ethic when you're not loaded.

PART 6

Sticky Situations in Beauty, Fashion, and Shopping

A killer blowout may not change the world, but it may change your attitude so much that *you* may change the world. Doesn't that sound so Zen?

You don't need to be a beauty editor to look your best all the time; you just need to think like one.

I used to marvel at a certain species of woman I lovingly call the fembot. You know, the girl who, on a humid tropical vacation, wakes up with sleek hair and perfectly applied eyeliner?

It took me years to realize that, nine times out of ten, Ms. Hawaiian Tropic has endured hours of thermal reconditioning and permanent makeup tattoos—not to mention credit card debt—to show up at breakfast looking ready for a reality show.

While I would never suggest you do the same, I think we can learn from this woman in her natural habitat. The key to feeling great is identifying which elements of your look are important to you and then being diligent about their maintenance while chilling out about the rest. If you live this way, the occasional beauty disaster will seem much more surmountable.

Think of yourself as a house. If the whole house is a mess, it's impossible to get started cleaning. But if just one room is in a state of disarray, you can get down on your hands and knees and scrub—or just close the door on the problem and refuse to let anyone in for a while!

In Beauty

You must look beyond fabulous. Immediately.

Think like a production executive: Only focus on areas that show. When you go on a tour of a film studio in Hollywood, you see that most of the "houses" used in films and TV shows are actually just facades—their exteriors are perfectly appointed, but open the door and all you see is scaffolding. When you have limited time, you must think of your looks the same way.

Now, I'm not suggesting you ignore all your covered-up places (though think of the cash you could save by skimping on bikini waxes!), but you must consider your audience when you need to look your best. If you're going to a party, chances are there's no need to perfect the area from your clavicles to your knees (unless you are planning to expose that arca, in which case I take back what I said about the bikini wax).

In general, I think one routine makes you look gorgeous for a night out, and another creates a seemingly effortless daytime glow. First off, moisturize. Nothing is uglier than dry, flaky skin. I've gotten addicted to Jergens Naturals Ultra Hydrating—it has cactus extracts and therefore doubles the moisture in your skin. I was a fancy-pants lotion addict until I found this at $6.99! At night, put your hair up and smooth your hairline with a wax product so you don't look like you're going to the gym. Apply sheer, shimmery eye shadow to your lids and blend it to your brow bone (L'Oréal has never let me down on this front). Apply mascara on the upper lashes only to avoid smudging. Check out RevitaLash—I may be seeing things, but it really *looks* like my lashes are longer. For a daytime run-in (with an ex, a crush, or the prettiest mom in your daughter's class): Take your hair down if it's up, and shake some volume into it. If the front looks messy, use your sunglasses as a headband for a sporty lady-of-leisure look. Keep a cute jacket in your car or a wrap in your handbag to cover up a ratty tank top or that men's thermal you wore to the gym. Daytime makeup can read trying too hard, so slick on some lip gloss (if you're willing to spend some bucks, check out Stila's; when I get cheap I'm a Maybelline girl myself) and run a tissue under your eyes to make sure no old eye makeup is giving you dark circles. If your skin is really sallow, keep a tube of Clarins Beauty Flash Balm in your bag—its old-school magic brings a flush to your cheeks. If you want a one-stop shop, go for Stila or Chanel. Neither one will ever let you down.

Be a Beauty MacGyver

It's not only MacGyver who can make magic out of the mundane—you can too. And there are no recipes for turning lawn mowers into hang gliders here—just some uncommon uses for common products you're sure to have around.

Vaseline makes a super-shiny lip gloss. It also smoothes crazy eyebrows.

Preparation H (yep, the hemorrhoid cream) will quickly deflate those bags under your eyes.

Hand lotion will tame that ungodly halo of frizz.

Olive oil, a shower cap, and a steamy soak in a hot tub will save you $50 on a salon conditioning treatment.

Baking soda will take the yellow cast off your teeth when you don't have time for a dentist trip.

Granulated sugar will exfoliate any areas prone to ingrown hairs and banish bumps.

Scrimp or splurge?

Some investment-price beauty products offer serious returns: an absurd number of compliments every time you wear them. Other items you can buy at the discount store and no one will know. Here are my two beauty shopping lists:

DRUGSTORE

Mascara

Treat yourself to a new $5 tube every month and you'll avoid eye infections and clumpy lashes. I like L'oréal.

Sunscreen

Dermatologists recommend basic, broad-spectrum, high-SPF drugstore brands because they're exhaustively tested, protect just as well as the fancy stuff, and are less likely to contain irritating fragrances.

False eyelashes

Unless you're Madonna, there's no need to spend a mint on mink.

Nail polish

Although they may not come in as many fashion-forward hues as those found in the nail salon, you can find the perfect pale pink for less than $5.

Lip gloss

Opt for the $25 fairy-dust-infused designer tubes and you're paying for an ad campaign. My secret weapon? The sweet, shiny candy glosses made for teen girls.

DEPARTMENT OR SPECIALTY STORE

Foundation

This is one item where service will offer you more than a smile—a consultant will help you choose the right shade and formula for you, and her advice is worth the price.

Someone wants to take your picture.

Know your angles. Do you have cheekbones that could cut glass? An amazing Roman nose? A little bit of extra flesh under your jawline? Pay attention to how the light will hit the areas you love—and the ones you hate. In general, light coming from above emphasizes the high planes of your face (the parts that stick out), and light coming from below emphasizes the well-padded parts. Hence, to minimize a double chin, position yourself below the light (better still, below the lens).

Fake complexion perfection. While a well-moisturized, dewy look appeals on the runway and on a night out, its charm morphs into grease on camera. Your best bet is to make everything matte with a sheer mineral powder, then dust a highlighting powder with a bit of sheen (look for products containing mica, an iridescent mineral) on your cheekbones, your brow bones, and the inner corners of your eyes. Also, be sure to carry purse-sized blotting papers.

Powder, eye shadow, and blush

Expensive powders are more finely milled than cheap ones, which means they disappear smoothly into your skin.

Fragrance

If you're a perfume girl, you know the difference between a delicate scent made from natural ingredients and a synthetic one that smells like bubble gum.

Eye cream

The pricey brands glide on most smoothly—a must when you don't want to tug on skin in the delicate eye area.

You appear to have aged twenty years—overnight.

First of all, have you been getting your beauty rest? I am amazed at how much better I look after eight hours of sleep. My eyes lose their puffiness and droop and my cheekbones somehow seem higher. Stress can also greatly affect the way you look. Getting a massage—during which the therapist never even touches your face—can work wonders, because your facial muscles will relax and exert less strain on your skin. If sleep and a massage are not realistic options, apply a rich moisturizer that will temporarily plump your skin and give it a dewy look. Choose one containing hyaluronic acid, which is a moisture-retaining agent naturally present in your skin. There's no need to spend tons of cash on a fancy department store cream—drugstore options will work just fine.

For a quick cosmetic fix, channel your inner French schoolgirl. Brush your hair straight back and secure with a plain headband, being sure to let the front pouf up a bit. This has the effect of a mini face-lift. Apply some sheer gel blush—I love the version in Jemma Kidd's line for Target—and slick on some clear lip gloss. Finally, curl your lashes (but resist the urge to layer on the mascara, as it will only make you look older) and be sure to bat them as frequently as possible.

If you're concerned about halting the aging process—and let me just go on record here as saying that I think there are far too many frozen-faced dolls roaming the streets and too few Meryl Streep and Susan Sarandon wannabes—you should visit a dermatologist for a consultation. No, Botox is not your only option. (Although in 2008, nearly 2.3 million people had Botox, and 14 percent were between the ages of nineteen and thirty-four.)

The dermatologist will examine your face for signs of aging such as fine lines, sun spots, white spots, redness, broken capillaries, and brown spots. (She should also perform a full-body exam to check for suspicious moles, since skin cancer is scarily common among young women.)

Depending on what's bothering you most, the dermatologist may prescribe a cream or in-office treatment. Retinoids—which are derived from vitamin A—boost collagen and control acne. Antioxidants fight free radicals that can lead to ag-

Practice your narcissism. There's a reason those girls on *America's Next Top Model* are always making faces in the mirror. We all have different smile styles that flatter us. For some, a toothy grin slims the face and neck, but on others, it can look chipmunky. Try looking at your face from a three-quarters point of view—this minimizes any asymmetry and brings attention to the eyes. And here's a secret tactic used by models worldwide to create alluring eyes: think of sex when looking at the camera.

Be body confident. There's nothing worse than a grown woman assuming the stance of an awkward preteen in a photo at a fancy event. Don't hunch your shoulders; this will make you look sheepish and sad. Throwing them back and keeping them down maximizes your chest and balances any extra width on your bottom half.

Put one foot in front of the other. Check out photos of starlets on the red carpet—they never stand with their feet parallel, but always cross one in front of the other to slim legs and give them an hourglass shape. The other tic celebs exhibit is putting their hands on their hips, which can feel a little forced for us mere mortals, but does a good job of narrowing upper arms and emphasizing the waist.

Think loooooong. Pull your shoulders down to create a swanlike neck; wear heels to add length to your legs; relax your hands like an expressive modern dancer instead of crunching them into fists like a prizefighter.

Let your hair frame your face. Photos are not the place to experiment with severe, slicked-back styles. If you have bangs, part them on the side and style them to float diagonally over your eyebrows to create a flirty look and camouflage any imperfections on your hairline or sun damage on your forehead. If you don't have bangs, style your hair toward your face and pull the ends over your shoulders. This will create the illusion of thicker hair and make your face look delicate in comparison.

ing. Peels exfoliate the top layer of skin to reveal a fresh one underneath. Laser treatments can treat a variety of issues and are usually prescribed in a series. Only your doctor can tell you which of these is right for you.

The one must for every single woman on the planet, whether she's concerned about aging or not? Sunscreen. As you know, it prevents wrinkles; even more important, it prevents cancer. You won't need to worry about looking old if you're dead.

You used acne meds to dry up your zits, but now your whole face is raw and peeling.

It's happened to everyone—you notice a breakout forming and, in an attempt to thwart it, you cover the offending area with teenage zit cream. The zit goes away, sure, but now your skin looks even worse than it did in the first place—big white flakes are dropping off, and it's raw and pink. For maximum damage control and to speed healing, use a washcloth soaked in cold water to gently cleanse the area. The washcloth will dislodge any flakes and let you see how the surface of your skin looks. If the skin is broken or near cracking, cover it with a thin layer of skin protectant (such as Aquaphor) and leave it alone until it starts to heal. If it's just dry, don't pick—simply apply a generous layer of unscented, gentle moisturizer such as Cetaphil lotion and leave it alone.

Dry patches like this can be really hard to

cover with makeup. What you need is a creamy concealer to stick to the dry spots and camouflage redness, and a translucent powder to seal it in. Apply the concealer with a soft brush, stippling it onto red bits. Use a light hand because you don't want it to cake. Then take a fluffy brush and lightly dust some powder over the area, blending it into the rest of the face. Next time, be gentle to your poor skin!

Your nose is always red.

Resembling Rudolph or Bozo does not make a girl feel her best. The way to cover up a red nose is to work some mineral powder foundation (Bare Escentuals makes an excellent one) into the area around your nostrils with a fine concealer brush, then use a fluffy brush to distribute excess powder up and across the rest of the nose. But coverage is just a quick fix—you need to determine what's making your nose red in the first place.

If you're sneezing and your nose is itchy—maybe you also suffer from watery eyes—you probably have allergies. Try taking an over-the-counter antihistamine such as Zyrtec or Claritin and see if the redness relents. If you find your nose reddens when you drink, you're experiencing a kind of allergic reaction. People who flush in response to alcohol—who are often Asian, hence the term "Asian flush"—lack an enzyme necessary to metabolize it. In which case it's best to lay off the sauce.

The cause of your red nose could also be rosacea—which, incidentally, is exacerbated by drinking. People with rosacea often have small broken blood vessels, pimples, or skin thickening in addition to redness (W. C. Fields' berry-colored cauliflower-shaped nose resulted from rosacea). Don't worry, though—although there is no bona fide cure for rosacea, many treatments are available. Dermatologists may prescribe antibiotics to get the condition under control from the inside out, as well as a topical cream or gel to treat surface skin. Regardless of whether you opt for a prescription, the National Rosacea Society recommends the following lifestyle guidelines to those struggling with the condition: Avoid alcohol and hot drinks. Stay out of excessively hot environments such as saunas and hot tubs. Avoid extreme weather. Limit stress, anxiety, and spicy foods. And since everyone falls off the wagon once in a while, it may be worth trying out an over-the-counter face cream formulated especially to treat redness. Many have a greenish tint that looks scary in the tube but magical on the face.

Sticky Note: The moment you feel a cold coming on, start religiously moisturizing the skin where your nose meets your upper lip. This will prevent the scaling and soreness that come from using too many tissues. Use oil-free products if you are prone to breakouts.

You know you should wear sunscreen, but it burns your skin or makes you break out.

Nice try, but you're not going to get out of wearing the stuff. Ask any dermatologist in the world what you should do to keep your skin healthy, and slathering your face in sunscreen will be the first thing the doctors insists upon. It's true that high-SPF sunscreens used to be irritating, because their level of protection required lots of chemicals. Now, though, companies have refined the sunscreen-crafting process to such an extent that formulas are both elegant—meaning they go on smoothly without chalkiness—and powerful. If you have extremely sensitive skin, look for physical sunscreens, which use minerals such as zinc oxide to provide a protective barrier.

You tried to get at a deep pimple—and now it's a big infected mess.

You know you're not supposed to perform surgery on yourselves, girls. Remember how this looks and feels the next time you have a big juicy one that's just longing for a good pop. (Sorry, that's gross—but don't lie and say you don't know what I mean.) Next time you feel a pimple coming on, prevent it from surfacing with this crazily amazing little heat machine called Zeno. It's expensive—around $150—but worth it, because it sends a targeted heat current right at the center of the zit to "kill" it. But back to the emergency at hand. Most dermatologists agree that you should leave the whole mess alone and make an appointment to see a doctor as soon as possible. You can calm redness by applying a cool compress, and if the area is tender and—there's really no other way to say it—oozing pus, apply an antibiotic ointment to keep bacteria at bay. Eruptions such as this are really hard to cover with makeup, so think about a using a small round Band-Aid and say you got a mole removed.

The Modern Girl's Ten Beauty Commandments to Prevent Nightmare Predicaments

Thou shalt wash thy face every evening.

Thou shalt moisturize and apply sunscreen every morning.

Thou shalt not iron thy hair more than twice a week, or color it more than once a month.

Thou shalt receive at least one pedicure every month.

Thou shalt apply no more than one shimmer product per look unless thou art under fourteen, in which case thou art too young for this book.

Thou shalt not wear a makeup color that does not occur in nature.

Thou shalt consume vast quantities of fruits, vegetables, and water.

Thou shalt move thy ass for thirty minutes at least five times per week.

You weren't supposed to lie down after Botox, but you did—for a loooooong nap. What to do now?

The reason most doctors tell patients not to recline after receiving their injections is to minimize the risk of the toxin—yes, it's a paralytic toxin—migrating beyond the muscles the dermatologist intended it to target. One of the most feared side effects of Botox treatments is ptosis, or droopy eyelids, which can occur when the drug goes to the wrong place. But it's rare. So if you've gone horizontal too soon after getting your fix, just sit up and hope for the best—there's very little chance anything will go wrong. If, worst case, it does, don't worry—the effects of Botox last just three to four months, during which time you could consider making gigantic sunglasses part of your signature style.

You've sprouted hairs in very unfeminine places: lip, chin, neck . . .

Every woman thinks they're the only one who has hair there. Not so much. Many of my girlfriends have confessed secret trips to the electrolysist, or simply shown up for lunch with watery eyes and a red, puffy upper lip. Excess hair is often no more than a consequence of being a va-va-va-voom Italian beauty, Greek goddess, or Latin lovely—the price one pays for full lips and bodacious hips. For many women of Mediterranean or Middle Eastern origin, hirsuteness is simply a fact of life that must be taken care of regularly, like gray roots or tartar. It's worth noting, though, that if you're the only gal in your clan with this issue, it may be worth checking in with your doctor to make sure your hormones are in balance. Sometimes something as simple as the birth control pill can make hairs vanish in a matter of months.

If everything hormonal seems to be on track, there are a number of surprisingly easy ways to get things under control. Laser hair removal uses heat to target hairs' roots, delivering them via a series of light pulses that feel like rubber-band snaps. It works best on women with fair skin and dark hair because the contrast is what enables the laser to target hair follicles. In order to achieve permanent hair reduction—no method offers permanent hair removal—a series of three to five laser treatments per area is usually necessary. Be warned, though: This is not a treatment for the meek. Use as much

Thou shalt remove any excess hairs darker than wheat and long enough to grasp with tweezers.

Thou shalt not gnaw upon thy nails unless thou art a contestant on *Survivor* and thou hast nothing else to eat.

numbing cream as you can lay your hands on, be brave, and chant "I will not look like Great-Aunt Gertie" over and over until it is finished.

Electrolysis is an old-school system that involves a technician inserting a skinny needle into hair follicles to damage them with heat. It hurts—a lot—but works on even the finest, lightest hairs. Expect to spend a lot of time being zapped, since every single hair has to be dealt with individually. And while it's not as bad as laser hair removal, don't forget you are getting an electric shock here, ladies.

For much less money, you can try waxing, threading, or sugaring, all of which work on the same principle—basically mass yanking. If you have sensitive skin, threading is probably your best bet, since the thread doesn't make contact with anything but the hair that's being removed. Incidentally, this is a fantastic method of eyebrow shaping, and you can usually get it done for less than the cost of waxing or tweezing.

This may sound a little Marcia Brady, but I have a friend (it really is a friend; I'm blondish and mostly hairless, but don't worry, I have other issues) who swears by surgical shavers in a pinch. If her upper lip is looking a bit shadowy, she just skims a skin-prep surgical razor (you can buy them online or at medical supply stores) over the offending hairs and they disappear magically. While this is a quick fix and not a long-term solution, it's ideal for those in the process of laser hair removal or electrolysis, both of which forbid you from tweezing between appointments. And since there's no lathering up, you don't really feel like you're shaving—more like buffing.

You have a red patch on your face that sort of resembles a breakout but isn't. It's red and itchy and a little bumpy, sometimes flaky, too. Nothing seems to help!

You'll need to see a dermatologist to confirm, but it sounds like you're being besieged by the ever-mysterious skin condition called eczema. Eczema is an autoimmune condition somehow related to allergies, and most people who experience it as adults had it as children. Eczema bumps often rear their ugly head(s) during times of stress and fatigue, and most people with eczema also tend to have itchy, dry skin. This is another one of those times I'm going to send you to the dermatologist, since it's virtually impossible to diagnose skin conditions without seeing them. However, I know that many of my friends who suffer from this mysterious affliction are eternally grateful for hydrocortisone, which calms inflammation. Whatever you do, don't treat eczema with acne products—these will only make matters exponentially worse.

If the rash appears mostly around your nose and mouth, it may be another strange dermatological affliction called perioral dermatitis. If you suspect you have this charming ailment—do a search to see scads of diagnostic pictures—do not, I repeat, do not apply over-the-counter cortisone cream. While doing so may cause the rash to retreat temporarily, it will actually make the problem worse in the long run. Perioral dermatitis shows up around your chin and puppet lines, and while there's no definitive consensus on what causes it, it affects mostly young women. If you have clusters of little red bumps that weep (such a sad and gross clinical term for oozing) and are a bit sore, definitely head to the doctor. Truly stubborn buggers won't vanish without a course of serious antibiotics.

You have a weird mole that seems to have developed overnight.

Go to the dermatologist—especially if you're fair, have a family history of skin cancer, or spend a lot of time in the sun. Moles that don't change over time and are symmetrical and uniform in color are not usually cause for concern, but the following are signs that a mole could be cancerous and deserves a professional look:

1. One side of the mole is different from the other.

2. The border of the mole is irregular.

3. The mole is a weird color or a combination of different colors.

4. The mole is bigger than the diameter of a pencil eraser.

5. The mole is raised above the skin's surface.

Any changes in a mole's appearance are cause for concern, especially if it bleeds, looks scaly, oozes, itches, or feels sore. Melanoma is the most common cancer in women between the ages of twenty-five and twenty-nine, and it's deadly—but it's also totally treatable, provided you get it looked at as soon as you notice it.

No matter which method you try in order to tame your bikini line, you suffer from angry ingrown hairs.

Ingrown hairs occur when hairs—especially curly ones—curve in on themselves and get trapped beneath the skin's outermost surface. Inflammation and infection cause the ingrown hair to resemble a pimple that can make a girl want to cancel her Hawaiian vacation. Prevent ingrowns by using an exfoliating lotion containing salicylic acid—the trusty yet unglamorous unisex Tend Skin is my fave—to free trapped skin cells after hair removal. If you're prone to this problem, avoid tight clothing and use laser hair removal or electrolysis instead of waxing or shaving. If you already have an ingrown hair, don't pick at it—just apply a warm compress to encourage the hair to break free, then top with

Tend Skin or a similar product. If you can see the top of the hair, it's okay to free it with tweezers—sterilized with alcohol, of course—provided you don't pull at the skin. If you draw blood or pus, dab at the area with a cotton ball soaked in hydrogen peroxide or rubbing alcohol, then apply an antibiotic ointment such as Bacitracin or Neosporin. A dusting of baby powder or a coat of calamine lotion can calm irritated areas you have no choice but to bare, as can lotions containing azulene, allantoin, and witch hazel. Just be sure not to use a product containing talc on your lady places—some studies connect it to increased rates of ovarian cancer.

You botched your self-tan.

I can't tell you how many times I've scrubbed my ankle bones raw in an effort to remove the brownish-orange blobs of discoloration from poorly applied self-tanner. But you don't need to remove an entire layer of skin to remove the color—there are products made precisely for this purpose (I love St. Tropez self-tan remover, available at Drugstore.com for about $17).

If you can't get your palms on a bottle of remover fast enough, there are plenty of homemade solutions. Massaging dry baking soda into wet skin in the shower will even out large blotchy areas. Cream facial hair bleach, applied to the palms of your hands and the soles of your feet for ten minutes, will remove any orangeness.

If your finger or toenails have turned terracotta, soak them in denture cleaner. And if you get a funny ring around your armpits—which probably occurs as a result of deodorant—just massage some baby oil into it in the shower, using a loofah or a washcloth.

However, this is one area of beauty where, like Sun-In for hair and baby oil for tanning, we've learned the hard way. You can make self-tanning easier. Try a more gradual tanner like Jergens Natural Glow Daily Moisturizer, which provides a more gradual tan, or a tinted formula that contains a dye to show you where you've deposited it.

You know you have to exercise, but you hate the postworkout hassle of showering and changing.

The eternal question: to hit the gym before or after the office? Go before work and you can schlep into spinning class in your jammies, then shower and get ready in the locker room. Work out after work, and you have the luxury of getting cleaned up in the privacy of your own bathroom—or lounging around all grubby while you eat fro-yo and watch reality TV. Those with reliable schedules can work in some midday fitness by doing a low-impact (non-sweat-inducing) exercise like yoga at lunch.

Whatever your schedule, it's worthwhile to invest in a gym locker so that you aren't forced to haul around a bag of damp track shoes and stretch pants. Cut down on your postworkout beauty routine by hitting the showers but skipping the shampoo. Just blow-dry around your hairline and use a drop of styling serum to tame flyaways. Wait half an hour after a rigorous workout before reapplying makeup. Skip the powder, which can cake, and instead use blotting sheets and top with a creamy foundation and cheek tint.

You got a really, really, really bad haircut.

First things first—how did this happen? You were sitting there the whole time, in front of a giant mirror, right? You weren't gagged and blindfolded and thrown onto some tragic makeover show? I'm not trying to rub salt in the wound, but you have to learn to direct every single beauty pro who puts his or her hands on you. Just because someone went to beauty school and does a really good blowout does not mean she knows what looks better on you than you do. Every time you sit down in a stylist's chair, you need to be able to clearly articulate what you'd like done, or the stylist's inner artiste may run wild and you may end up looking like a runway runaway.

That said, sometimes getting a bad haircut is like watching a car crash—even though it seems to be going by in slo-mo, you just can't stop it.

So consider this list of disaster-aid tips the hair equivalent of the jaws of life.

- **Dry shampoo is your friend.** The whole concept is a little weird—after all, Psst, the original dry shampoo, was invented for hospital patients who couldn't make it to the shower—but this spray powder adds great texture to hair. Just spray it into your roots and flat hair will get an instant lift. It also makes short bangs smooth-back-able.

- **Invest in lots and lots of bobby pins.** Twist and pin up crazy bits in a haphazardly sexy, European way.

- **Pull your hair back.** If your ponytail is depressingly stubby, here's a trick that will make it seem to have sprouted inches overnight. Part your hair horizontally from the top of one ear to the top of the other—so you've now divided your hair into two sections. Pull the top section into a ponytail centered at the back of your head. Now take the rest of the hair and pull it into a low ponytail just below the first one. Then pull them together in one hair band. Voilà—one giant pony. It will take a while to master this style, but trust me, you'll turn to it constantly.

- **Flat-iron it.** This doesn't work for all styles or hair types, but it makes hair seem longer.

- **Prenatal vitamins.** Some people swear that prenatal vitamins speed hair growth. There's no evidence this is true, but they can't hurt, right? Remember to ask your doctor before starting any supplement plan.

- **Consider going to another, more trusted stylist for a consultation.** Do not let this stylist cut! Just ask what he would do to improve the look of your new hated do. Heck, chances are it couldn't be worse, right?

Sticky Note: Not sure what to tip at a hair salon? Between 15 and 20 percent of the bill is customary, excluding any products your stylist suggests from the total. If you have more than one stylist, the total tip you leave should still equal 15 to 20 percent; just split it up proportionally. Even though many salons will allow you to leave a tip on your credit card, most stylists will appreciate the cash immensely, so plan ahead and hit the ATM. You do not need to tip the owner of the salon if he or she cuts your hair.

Your highlights are more Raggedy Ann than Barbie.

Brassy is the word the pros use to describe hair color with too much of a red tone, and this is a common affliction for brunettes who get highlights. If your hair is darker than medium brown, it has to pass through a reddish stage before it will hit blond. Many inexperienced stylists are afraid to let hair color keep lifting past the red stage, for fear of damage, but it's necessary to achieve a buttery, golden look.

If you notice the brassiness while you're still at the salon, by all means mention it to your stylist and ask to have it corrected. The stylist can either relift your highlights or apply a toner that will play down red undertones.

If you don't notice until you get home, call the salon and ask for a corrective appointment. The salon should fix, free of charge, any color you're not happy with; at the very most, you may be asked to cover the cost of the chemicals used.

In the meantime, pick up one of the shampoos for silver and white hair—they're not just for grannies anymore.

Sticky Note: Can you go back to your hairstylist once you've cheated and gone somewhere else? While it may feel this way, you are not actually wedded to your hairstylist . . . or waxer, manicurist, gynecologist, or any other service provider. If you "strayed" and tried someone else but decided you prefer your original provider, hold your head high and head back in. Be honest and say you were trying something different (or you can fudge and say a friend gave you a gift certificate—though that's a little less helpful at the gyno), but you missed your old friend and his or her great ways. Trust me: People are sorry to see customers leave. They are rarely sorry to have them back.

You're afraid to say it out loud, because then it's really true: Your hair seems to be thinning.

Like chin hair and stretch marks, this is a feminine affliction many women think they're alone in. Please, girls—do you really think you're so special? Not! According to experts, more than 30 percent of women will experience hair loss by the time they're fifty. More of your friends than you think start the day with scalp serum, thickening spray, and a dose of female Rogaine.

The first thing you need to do is make an appointment with a dermatologist who specializes in alopecia, or hair loss. He or she will analyze your scalp tissue and diagnose the problem. You may need to get some blood tests done, to check for hormonal imbalances or vitamin deficiencies.

Once you're under a doctor's care, there are lots of tips and tricks to camouflage the problem. Get regular haircuts, so that the ends of your hair appear blunt and healthy. A style that's shoulder-length or shorter does the most to flatter less-than-lustrous locks. Apply an amplifying spray or lotion, and blow-dry with a round brush to create lift at the root. If you have curly hair, dry it with a diffuser while you're bending over.

And don't be afraid to wash and brush your hair. This will not make you shed more, contrary to popular belief. It may stimulate blood flow to the scalp, though, which is a good thing as far as those sleepy follicles are concerned.

Sticky Note: Although extensions are a boon to women whose hair is fine or refuses to grow past their shoulders—just ask anyone in Hollywood—having them glued or sewn in repeatedly can actually weaken hair follicles. Be sure to give your hair a break between applications, and consider clip-in extensions if damage seems to be an issue. (Look for ones made from human hair.)

The Ultimate Sticky Situation Emergency Kit

- One breath mint
- Hairspray
- Double-stick tape
- Two Tylenol
- Tampon
- One Band-Aid
- Nail polish
- Wet Ones wipe
- Thong
- Oil blotter
- Lip balm
- Needle with thread
- Matchbook or lighter
- Nail file
- Elastics
- Mirror

In Fashion

Fashion emergencies are at least as common as bad hair days, but they don't seem to get as much sympathy. Why? I think it's because while you can't help the genetic code bestowed upon your unruly tresses, a fashion emergency is almost always your fault. People think, and rightly so, that you could have prevented it.

You should take ownership of your style choices, but sometimes your outfit just falls apart—literally—and you can't help it. There's no wardrobe malfunction I haven't experienced. My shoes have fallen apart, my bra straps have slid off, my hems have fallen, my buttons have popped open, my seams have split. Long before Janet Jackson let her nip slip, I filmed an entire intro for my show on the Style network before anyone told me my entire left breast had been exposed like that of an Amazon on an X-rated version of *Wonder Woman*.

Anytime you're the last to know about a fashion gaffe, all you can do is laugh it off, even when you're cringing inside. Just make a joke and move on, realizing that everyone has seen somebody else naked, and that everyone owns at least one ridiculously heinous item she thought was a good idea at the time. The more you apologize, the more you talk about how embarrassed you are, the more attention you will draw to the incident. Apologize or comment once. Then move on.

Matters of Taste

You need the perfect interview outfit but don't have time to shop.

You may not realize this, but it's your lucky day—the worst thing to do before an interview is shop for a new outfit. You want to wear something you're already comfortable in—a skirt whose hem you know doesn't ride up, a blouse you're 100 percent positive isn't see-through. The last wormhole you want your mind wandering into during a job interview is the "I look so disgusting/unstylish/fat/silly/overdressed/underdressed/juvenile/geriatric" self-esteem vortex. Especially when you're supposed to be coming up with the world's only unique answer to "What's your biggest weakness?"

That said, what they say about second chances and first impressions is true, and there's no denying that our society judges books—and writers—by their covers. So, what can you find in your closet that will convince your potential boss that you're perfectly dressed for the job the company has—and which you want?

First, close your eyes and imagine your ideal self doing the job you're going for. What's this best version of you wearing? Okay, if it's a custom Dolce & Gabbana tuxedo, you may be out of luck in recreating it. But this fantasy image will help you see what your confident self looks like. And while the validity of all that stuff in *The Secret* is debatable, a little visualization never hurt a pro athlete.

I'm of the opinion that it's okay to dress just a little bit too nicely when you're trying to impress someone. In the years since I've been interviewing candidates to work with me, the office has become a more and more casual place. But even if the CEO wears ripped jeans to work, it's never okay to show up in flip-flops. Especially at an interview.

While every office environment dictates different fashion rules, there are a few hard-and-fast ones that apply everywhere:

- **Comb your hair.** Bedhead looks good on billboards—and in bed, obviously—but when you're trying to get someone to give you money, you want that person to be confident you know how to operate an alarm clock.

- **Check your nails.** By no means do you need a pro manicure—though that's always a nice touch—but you should not have dirt trapped under your fingertips, unless you're a mechanic or gardener (in which case the presence of dirt might prove you're a hard worker). You'd also do well to remove any chipped nail polish and file any snags—you don't want to scratch your interviewer during that confident and powerful handshake. And while long red talons may do nicely when grabbing a man, they can make a different impression at a job interview. Your employer may be wondering whether you will be spending more time worrying about whether you might chip a nail than getting the hard work done.

- **Keep your underwear under wraps.** Leave lingerie chic to nineties Madonna and cover up your lady bits. This means no low-cut necklines, high-cut minis, or anything see-through. It's probably a good idea to keep lace off your outerwear as well.

- **Fit is everything.** Whether it's a skirt, jacket, or pair of boots, nothing should pull or gap. If you own foundation garments, wear them—Spanx are not just for under fancy dresses. The confidence you'll gain from knowing nothing is lumpy, bumpy, or protruding beyond where it's supposed to will show in your swagger.

- **Wear shoes you can walk—even run—in.** Nothing screams "incompetent narcissist" more than a brutally painful pair of stripper shoes. That said, if your stilettos give you confidence, by all means wear them—just keep the rest of your outfit a bit subdued. But no sneakers or flip-flops, please!

- **Really pay attention to what kind of job you're applying for.** You can even ask the receptionist or whoever's interviewing you what the atmosphere is like. Many fields, such as banking and law, require a bona fide, grown-up suit. But if you're in search of a job in a creative field—such as graphic design, fashion, or publishing—you can get more, well, creative in your apparel. A fashion-forward employer requires a fashion-forward outfit; you might even consider borrowing a status bag from a friend in order to dress to impress.

- **Shoot to look just a notch better than your prospective coworkers.** Err on the side of too formal—it doesn't hurt to let them know that you want the job you are applying for. It's flattering.
- **Natural beauty is the goal.** No overbearing perfume or too much makeup.
- **Most important, don't forget to be yourself.** If you need to raid the costume box in order to prep for your interview, the job's not for you.

You arrive at a party in your flawless new dress only to spot someone else in the exact same sheath.

Sigh. All you can do is embrace the coincidence. Make an immediate beeline for your unintentional twin and compliment her on her fabulous taste, then smile big and pose for a photo. And look on the bright side—maybe this stylish girl is your brand-new partner in crime, or at least your closet-shopping soul mate.

You are decked out in your finest, while other party guests are strictly casual.

Feeling self-conscious? Cover up any bared skin with a jacket or pashmina, let your hair down—literally—and remove any too-flashy jewelry. Not interested in hiding your hotness? Work your look and hint about the far more fabulous soirée you are party-hopping your way toward.

You get to the party and are way underdressed.

At least you don't look as though you're trying too hard. Give your outfit a quick assessment—pull off your hoodie, tuck in your T-shirt, and neaten anything that can be neatened. Take your hair down from a casual ponytail, or pull it up into a quick French twist. Add some extra makeup—powder and mascara do wonders. Most important, stand tall and walk with confidence. You may not be the best-dressed girl in the room, but at least you are the most comfortable. Don't sabotage a potentially fun evening by spending it apologizing for your outfit. And by the way, if you're in Los Angeles, you may have done yourself a favor by underdressing. In LA there seems to be a bizarre inverse relationship between the level of a person's fame and how disheveled he or she is.

You didn't know it was a costume party.

Your first line of defense is to ask the host for help—often she'll have a stash of wigs or silly glasses on hand for clueless or reluctant guests. If not, hit the bathroom with any makeup you have—use your eyeliner to draw on faux bruises or to give the appearance of ghoulishly hollowed cheeks and claim to be a dead celebrity. Another simple option: Wedge a throw pillow under your dress and call yourself Angelina Jolie.

You spent the night with someone new and now you don't have enough time to run home and change before work.

We've all done the Walk of Shame. Your nosy coworkers will notice if you repeat your top or dress from the day before, so your best bet is to layer over it. Can you borrow a T-shirt or button-front from your new paramour to wear with yesterday's skirt? Do you have a shawl that can be layered over your dress or tied into a makeshift top? Once you've mixed up your outfit, hit the drugstore for a coif-reviving can of dry shampoo and a fresh pair of tights (perhaps in a contrasting color). Spray down your hair and give it a quick shake before twisting it into a simple bun, then let your lipstick double as a cream blush.

You need chic and easy travel duds that fit into a carry-on.

When you're traveling with someone who refuses to check bags, you must pack lean and clean. Layering is the key to a comfortable and compact travel wardrobe. Start with a base of leggings and a long-sleeved tee (or tank, if you are traveling tropical) and add to it. Tissue-weight sweaters can be worn solo or in multiples for extra warmth. A convertible jersey dress serves as a strapless in the evening and a skirt during the day. Dress up your simple separates with a colorful shawl (which doubles as a handy flight blanket) and a few pieces of statement jewelry. And don't limit yourself to boring black shoes—bring along a wild (but comfortable) pair in, say, purple patent and make them your signature accessory. Superstylist Rachel Zoe uses her oversize sunglasses as a sleep mask on the plane or a headband after dark.

You're traveling abroad and don't want to look like an accidental tourist.

Part of the fun of travel is getting inspired by local street style and discovering new trends and designers. Bring sleek, easy travel separates like leggings, wrap dresses, and tissue-weight sweaters and add a pair of comfortable, high-style walking shoes such as knee-high flat boots, patent wedges, or fun, colorful booties. Top it all off with a shrunken leather jacket and a pashmina shawl, then get ready to accessorize with cool finds bought on location.

You're going out with your hipster/banker/fashionista friends and want to fit in with their crowd.

The most important thing is to stay true to your own style. Wear something appropriate for the formality of the venue or event, but don't pose or bend over backward trying to fit in. When hanging out with hipsters, the key is fearlessness—if you own what you're wearing, you'll command respect. Finance people like quality and subtlety. If you want to be taken seriously in the banking crowd, choose a simple outfit that will highlight a prized piece, whether it's a vintage cuff or authentic riding boots. And fashionistas feel anxious around other fashionistas, so unless you have a subscription to Italian *Vogue*, don't even try to compete. Just sit back and enjoy the fashion show.

Fashion scrimp or splurge?

You know that girl who always seems to look as though she's leapt from the pages of this season's *Vogue*, even though you know she doesn't make a dime more than you do? How does she afford to deck herself out in current styles and still stay current on her rent? It's the same strategy that keeps interior designers in business and little kids in socks: the art of mix and match. While it pays to save up for iconic pieces you'll want to give your daughter one day, don't bother shelling out tons of cash for a flash in the pan or a piece you're going to hide underneath other ones. Here are two shopping lists: one for that massive mall store, the other for that exclusive designer emporium.

Big Box Store (Target, Walmart, and so on)

Cotton basics

These include T-shirts, panties, socks. If they feel stiff at first, just run them through the wash a couple of times and go heavy on the softener.

Hosiery

The best black tights I've ever bought were $5.99 at Target.

Flip-flops

There's often no difference between the fancy ones that cost $24 and the inexpensive ones that cost $4—except you can buy six times as many.

Jeans

Okay, I know I am speaking against the fashion-girl gospel ("Thou shalt covet $500 bottom boosters made from the finest in Japanese cotton"), but the truth is, discount stores have finally gotten hip to the designer dungaree trend, and they're making some amazing pairs. Shhh: The juniors' department is a treasure trove.

Woven Shirts

From hippie pullovers to striped oxford-cloth button-downs.

Jewel Box Store

Dresses

Whether it's an everyday LBD or a one-night-only sequin extravaganza, it pays to pay for a frock that fits perfectly and whose fabric will make you shine with confidence, not sweat (read: no polyester).

Bathing Suits and Bras

If you were building a house from scratch, you wouldn't want to hire anything but the best architect. Underpinnings are works of engineering, too—so make sure your structure doesn't collapse. Spend some cash on high-tech fabrics and a perfect fit.

Coats

This is the best example of cost-per-wear of any item in your closet. When you consider how frequently you're going to throw on that trench, you're not likely to suffer from sticker shock.

Handbags and High Heels

The dynamic duo of dressing rich will keep you in fashion no matter where you end up. In Paris and Tokyo, you see girls in their boyfriends' clothes just like the ones you wear for Sunday brunch in the United States, but because they've punched up their look with high-profile accessories, they look like fashion models, not off-duty cheerleaders.

Sticky Note: When it comes to design and affordability, it doesn't always have to be either/or. In recent years, stores like Target have become patrons of some of the world's most talented designers, allowing them to offer their creations at a fraction of what they'd cost in their own boutiques. Stay current on special designer offerings by keeping up to date on your favorite stores' Web sites.

Faux Real?

Costume jewelry—which simply means that it's not made of precious stones—has been chic since Coco Chanel reinvented it for the fashion crowd almost a hundred years ago, but sometimes you want the genuine article. Just as with clothing, sometimes it pays to spend—and sometimes it doesn't. Here's what you'll find in my jewelry box.

Faux diamond studs, but real gold hoops

Faux mega-stone cocktail rings, but a real diamond engagement ring

A tangle of long faux gold chains, but one delicate platinum and sapphire pendant

It may seem counterintuitive, but in general, it's best to spend big on small items.

You're meeting your man's family for the first time, and you don't know what to wear.

Take a deep breath—he loves you, so why shouldn't his family, right? If only it were that simple. First, ask your boyfriend if his family has any religious or cultural differences you should be respectful of. It's always best to err on the side of too conservative. Skip the stilettos and monster trends and choose something pretty and feminine, like a 1940s-inspired floral dress. Top it off with a cardigan you can remove with ease—you don't want to be sweating at the dinner table in a too-hot pullover.

You're meeting your boyfriend's friends for the first time, and you don't know what to wear.

It's okay to be a little sexy, but don't sacrifice comfort. If things go as planned, you'll be seeing a lot of this crew, and you don't want to have to put on a persona—or a push-up bra—every time you join them for happy hour. Choose a tried-and-true outfit that you feel confident wearing. This is the time to break out the killer jeans and loose-but-maybe-a-teensy-bit-suggestive top. Your man is showing you off, after all—and most men care more about the woman wearing the clothes than the clothes themselves.

You can't seem to stop hauling around an overstuffed handbag.

Many of us have turned our bags into a sort of security blanket, hauling around an ever-expanding arsenal of things to address any setbacks that might come our way. The problem is, we rarely use the spare slingbacks and mini umbrellas that are adding up to backaches. So, let's look at the contents of your purse element by element.

- **Wallet.** Start by replacing your heavy leather wallet with a lightweight nylon number that holds only what you need: cash, credit cards, and ID. Jettison the checkbook, the photographs (you can upload the latest shot of your Chihuahua to your iPhone), and all the preferred-customer cards (switching out to the keychain-sized versions for the one or two stores you actually frequent).
- **Cosmetic bag.** Select a few products that can do double duty, like cheek and lip tints or powder-finish concealer, and dump anything that is tricky to apply on the

go, such as foundation or liquid eyeliner. A simple pencil liner with a built-in smudge sponge can immediately sex up a daytime makeup look and won't take up much real estate.

- **Keychain.** Unless you're a janitor, you shouldn't be dragging around dozens of keys. Keys to your friends' apartments (left with you for safekeeping) or your parents' house (which you visit twice a year) should go in a central spot at home. Test out the rest of them—are you carrying around your brother's ex-girlfriend's mailbox key without knowing it? Pare down to only the necessities: keys to your house, car, and office. And while you are at it, swap out your heavy monogram keychain for a simple (and lightweight) ring.
- **Emergency supplies.** Boy Scouts are known to be prepared, but those badge collectors have nothing on us. If you don't want to go without tampons, Band-Aids, and hand sanitizer, buy a small zippered bag and fill it with mini versions of your favorite necessities. Disposable teeth wipes, applicator-free tampons, and a tiny tube of concentrated hand lotion will keep your backup plan in place without weighing you down.
- **Tech stuff.** These days, there's absolutely no reason to be carrying around a separate phone, schedule, and digital camera. Do your research and invest in a single device that will handle all your tech needs. Pay special attention to the length of the battery charge—you don't want to have to carry around a charger.

Wardrobe Malfunctions

Your slingbacks won't stay on.

Slap, slap, slap—never the sound you want to hear from your shoes once you've upgraded from flip-flops. First, make sure that your shoes are the correct size. You probably haven't had a proper fitting since your last teenage growth spurt, but it never hurts to double-check. Feet can change size and different manufacturers have different standards. Always buy the size that fits you, not the one with the "right" number on it.

If your shoes fit but your slingbacks are old, the elastic might be loose or the leather stretched out. Straps can be shortened at a shoe repair shop for a better fit. You can also try painting the back of the strap with clear nail polish, allowing it to dry, and then rubbing it with a nail file to make the surface a little bit sticky, so it will adhere to the back of your foot.

Your bra wire is popping out and stabbing your boob.

Ouch! First, you need a quick fix. If you're ready to sacrifice the bra to the lingerie gods, just pull out the underwires and go without for the day. If you can't do without the extra support, tape them—the wires, not "the girls"—into your bra with duct tape or bandage tape. You can make a more permanent repair by hand-stitching the wires into place.

Underwires pop when a gal wears the wrong band size. If a bra doesn't fit snugly around your torso, it will rub against the body, wearing away the casing that houses the wire and eventually allowing it to rip through. Try to get a professional fitting if you suspect your band size might have changed as a result of any fluctuations in weight.

Lazy laundering can also cause shapers to break through. It's so tempting just to toss your dainties in the machine, but stretch fibers cannot withstand heat, which causes them to break down and robs the garment of its stretch. If you must machine-wash, use cold water, on the gentle cycle, and secure all lingerie in net bags. Always air-dry.

You have a snag in your sweater.

You brush your sleeve against your necklace and snag a thread—one of those moments when you wish you could travel back in time. Ignore the temptation to pull on the loop or chop it off with scissors. Instead, use a crochet hook to draw it through to the inside of the sweater. Then turn the sweater inside out. For extra protection, use the hook to make a knot on the inside of the sweater, then daub the end of the yarn with a ravel protector (called Fray Check) or clear nail polish. Let it dry before turning the sweater right side out.

Your zipper is stubborn.

Whatever you do, don't try to force it open or closed. First, relieve the pressure on the zipper (this is why eighties ladies went horizontal to zip up their jeans, but I don't recommend it—nor jeans that require it). Then rub a crayon against the jammed section of the zipper; the wax will help its teeth slide open.

Replacing a zipper can be a complicated proposition—it's a job best (and most cheaply) handled by a professional tailor. If you have to force your zipper closed every time you wear a garment, consider having a few seams let out to ease up on the fastening drama. Or buy a bigger size.

Your clothing is covered in pet hair.

If you're en route to a party looking like Chewbacca, defuzz yourself with a big loop of masking tape wrapped sticky side out around your fingers. If you can't find any tape, put on a dishwashing glove and rub it back and forth against the errant pet hair—clumps will roll into easy-to-remove strands.

If you have a shed-happy pet, you should keep lint brushes (or rollers) stashed at home, work, and the gym for frequent cleanup. Research the best vacuum cleaner for your pet's breed and use it frequently, especially if you have loved ones with allergies. To cut down on stray hairs, brush pets with a shedding device like the Furminator, which removes the dead undercoat of animal hair (you will be shocked and awed—and maybe horrified—when you see how much hair is lurking beneath the surface). Special pet shampoos and conditioners that reduce shedding are also available—ask your vet for details.

Your sunglasses don't fit—or do they? You're not sure.

When choosing a flattering pair of sunnies, keep your face shape in mind. A round face looks best in rectangular or square frames that will counteract curves, while those with a square jaw line might choose rounded Jackie O frames for a softer look. A heart-shaped face is complemented by narrow aviator shades that will deemphasize a wide forehead, while lucky girls with oval faces can play up their symmetrical features with nearly any frames.

Choose a shape that covers your eyebrows but follows their general line. Look in the mirror and smile big—you want to make sure that the frames do not ride up the bridge of your nose or rise with your cheeks as you smile. Shake your head to ensure that they aren't too loose, and finally check that your eyes are centered in the middle of the lenses.

Though cheapie shades are a great way to experiment with different looks, purchasing glasses from an optometrist means you will benefit from a professional fitting. (Not to mention the fact that you'll be assured the lenses are high-quality—remember, above all else, sunglasses are meant to protect your eyes!) To cut costs on high-end frames, try them on at the eye doctor and note the model number, then purchase them online at a fraction of the price. Another high-fashion, low-rent option? Buy a pair of vintage prescription glasses on eBay for a song, then have the lenses replaced with your own Rx or plain tinted plastic. Be sure to get a binding estimate before having lenses custom-made, though, as prices vary wildly from shop to shop.

You have a run in your tights—and no time to change them.

Your mom—or grandma—would tell you to dot some clear nail polish on the snag ASAP to prevent it from running further (a spritz of hairspray also works). Prevent tears in the first place by checking your nails for snags before putting on your tights and avoiding shoes with buckles—or boots with zippers—that may pull on fragile nylon.

If you're marooned in a pair of laddered tights, work them! Cutting-edge designers show busted, punk-inspired hosiery on the runway, so you can make it a fashion statement in real life. Or make sure you're well moisturized enough to go bare-legged at a moment's notice.

All your shirts have sweat stains.

Start with the source. If you're a profuse sweater, tell your doctor, who may prescribe a heavy-duty antiperspirant. (If you think anxiety may be the cause of your wetness, be sure to let your doctor know—other medications may work best for you.) Believe it or not, Botox injections can also stop perspiration by paralyzing sweat glands.

To remove existing sweat stains from clothing, make a paste of baking soda and water and let it sit on the stain for twenty minutes before rinsing, then launder as usual. If that doesn't work, try soaking the garments in plain white vinegar.

You can prevent sweat stains by having fabric sweat guards (also called dress shields) sewn into your clothes by a tailor, or sticking on a disposable version. And of course it never hurts to embrace dark colors.

Your hangers leave dimples in your clothes.

It turns out that Joan Crawford in *Mommie Dearest* was right when she admonished her daughter with the edict "No wire hangers ever!" Dry-cleaner freebies are murder on clothes and best used for breaking into cars, toasting marshmallows, and building abstract sculptures. Invest in wooden or fabric-covered hangers—both Ikea and Target offer excellent options—that will maintain the integrity of your garment. Add stick-on felt circles to each side of the hanger to keep delicate straps from sliding off.

Since knit fabrics such as cotton jersey, wool, and cashmere are easily pulled out of shape, it's better to fold them and keep them on shelves or in drawers than to store them on hangers. Certain domestically gifted queens of all media like to fold their sweaters around layers of acid-free tissue paper, but I wouldn't even know where to buy such a thing. I prefer to roll knits to save space and avoid wrinkles.

If you must hang your knitwear, use a padded hanger or fold a sweater in half lengthwise and hang it over the bottom bar of the hanger like a pair of pants. To avoid creases, let garments cool completely from the dryer before hanging or folding.

Your jewelry is dull.

The best way to keep your jewelry in its best shape is to store it carefully. Velvet- or felt-lined boxes will cushion your favorite pieces while protecting them from dust and dirt. If your pieces need a good cleanup, be sure to treat them properly, according to what they're made of. There's no need to buy overpriced jewelry cleaning solutions, though—just play kitchen chemist and whip up one of the potions below.

- **Sterling silver.** Place the piece on a sheet of aluminum foil in a heat-proof pan. Pour in some boiling water and add 2 tablespoons of baking soda and 1 tablespoon of salt. Let the jewelry soak for a few minutes, then carefully remove it with tongs. Rinse in cold water and buff with a chamois rag.

- **Gold.** Soak jewelry in a solution of 1 teaspoon ammonia and 1 teaspoon of dishwashing liquid dissolved in 1 cup of warm water. Remove after twenty minutes or so and clean with a soft brush.

- **Gemstones.** Soak in vodka, then gently brush with a toothbrush. This is a great way to use up the cheap vodka you didn't manage to sneak into the punch at your last party.

- **Pearls.** Dampen a soft cloth with olive oil and gently buff.

- **Costume jewelry.** Avoid solvents since you can't be sure what the piece is made of. Clean with warm water and a soft toothbrush.

Your hem fell out of your pants.

What else did you think that stapler on your desk is for? Just fold pants according to their permanent crease, then anchor the temporary hem in place by stapling from inside cuffs, so the staple ends don't catch on your skin or hosiery. This is obviously a temporary, and imperfect, solution—but it will get you through the day. You can also repair a hem yourself using fusion tape, which you can buy at any craft, fabric, or discount store. Cut a piece to match your hem's droop, slip the tape into the fold of the hem, and follow the ironing instructions. You can also try magnetic clips, like those made by Zakkerz. They keep rolled-up cuffs crisp-looking, perfect for wearing flats with a pair of extra-long trousers.

You're not sure whether "dry-clean only" really means dry-clean only.

In a word, not really. Okay, that's two words, but you catch my drift. Manufacturers use these labels primarily to cover themselves from being blamed for damage inflicted on their garments by machine washing. To be safe, I dry-clean garments that say "dry-clean only," but not those that say simply "dry-clean." But many of my friends say I am wasting money, and that pretty much anything can be hand-washed.

Here's why: There's nothing dry about dry-cleaning. Garments that could potentially be damaged by water are instead dipped in chemical solvents. While most wool items, metallic blends, and structured garments with boning or built-in shoulder pads should be cleaned by this method, most other items can be washed by hand, or even in your machine on the gentle cycle. Invest in oversized netting bags (usually meant for lingerie) to protect delicate or embellished pieces in the machine, and air-dry anything with stretch to it. You can also try one of the new at-home dry-cleaning products, which steam-clean your clothes in the dryer. Not only is this system an affordable compromise, it can help you avoid some of the intense chemicals that spring forth from the plastic bag when you open your dry cleaning.

Ultimately, the most useful service a dry cleaner provides is its professional pressing. It may be worth paying a few extra bucks to avoid hours slaving over your ironing board.

You hit the streets in your new micromini only to realize that it's a little shorter than you're comfortable with.

Uh-oh! Is it possible that your new dress is actually an—um—top? If even an eighth of an inch of your bottom is showing, you might want to keep your coat on, or run to the drugstore for a pair of opaque tights. Ultimately, though, these situations are all about attitude. Acting self-conscious and awkward will only draw attention to your tush—and your discomfort. Channel the fashion confidence of a TMZ starlet—just for one night, like Cinderella—and welcome any lucky bystanders to feast their eyes.

Your brand-new shoes are slipping all over the place.

Before debuting a new pair of pumps, it's best to rough up their soles with a piece of sandpaper, or score them with an X-acto knife. Can't bring yourself to ruin the red lacquer soles of your Louboutins? Slap on a square of heavy-duty bandage tape or duct tape. Or try what wardrobe stylists use backstage at fashion week—spray on some sticky hairspray or dribble on a little sugary soda for a bit of grip. I avoid this problem by having a shoemaker glue rubber soles to my shoes before I wear them. And for egoists like me, they now even have red soles!

You get caught in the rain in your favorite piece of suede.

If your coat is in peril, just turn it inside out—you might look a little goofy, but you'll spare the leather.

If it's your bag you're worried about, slip into any retailer and ask for a plastic shopping bag to cloak your nonplastic one in.

If your boots get soaked, park them next to the radiator or space heater and let them dry completely, then really work them with a suede brush. You can also try buffing off stubborn stains with a fine-grained emery board.

Try to get your suede goods waterproofed as

soon as you buy them—many shoe and department stores will spray down items for free if you ask. Otherwise, pick up a silicone-based waterproofing spray at the drugstore or shoe repair shop.

You always get baggy knees when you wear skinny jeans.

Ah, the curse of the elephant knees! If you plan on wearing your skinnies with boots, try pleating the excess cuff material and then rolling it up tight. Okay, yes—I am actually suggesting you peg your pants. Don't worry, though—this über–1980s fold will be covered up by your boots. You can also tuck cuffs into a heavy pair of knee socks. Or, for a sleeker look, buy a pair of equestrian pant clips. These nifty (and cheap) pieces of elastic clip onto either side of your cuffs, which are then pulled tight under your foot.

Another option is to buy skinny jeans with the new X-fit Lycra. Stretchy fibers are woven into both the warp and weft of the fabric for a four-way stretch that retains its shape.

Your purse zipper breaks while you're out on the town.

Suddenly there you are with all your valuables (not to mention your crumpled receipts and used tissues) exposed to the world. First, zip your money and credit cards into an inside pocket. Next, secure any extra fastenings (such as latches or magnetic clasps); hold your bag under your arm, clutch-style, and clamp it closed with your elbow until you can offload the busted bag.

Most broken zippers can be mended or replaced by a cobbler or leather repair shop. Call the store where you purchased the purse and ask whom they recommend to repair their products. Often, high-end bag companies have their own mail-in repair services to ensure a perfect fix. Just keep in mind that such services can be quite pricey; always ask for a quote before sending off your pocketbook.

Your hot knee-high boots are too big (or too small) in the shaft.

Few women realize that boots can be tailored at a shoe repair shop to ensure an impeccable fit. Have a pair that gaps around your legs taken in—just be sure to leave enough space for whatever hosiery or jeans you want to wear underneath. Too-tight boots can be stretched by your cobbler to accommodate muscular calves.

You love heels, but they kill your feet. Also, you can't walk in them.

Lots of traditional comfort brands like Aerosoles and Easy Spirit have revamped their lines to include fashion-forward shoes that are easy on the soles. Cole Haan has red-carpet-ready platforms with concealed Nike Air pads. Or try ballroom dance shoes; they have vintage appeal and are made for movement. Make your prettiest pairs easier to walk in by having them stretched at the shoe repair shop. Or add stick-on pads where the shoes rub against your skin.

The more you wear heels, the sooner you'll be comfortable in them. Wear your heels around the house to build up your confidence. Take small steps, putting down your heels first and then your toes. Take your first venture out wearing lower heels; work up to high wedges and finally stilettos. Watch *America's Next Top Model* for inspiration—and comic relief.

Another option is to take your high heels to the best shoemaker you can find to see if he can change the pitch—some shoes can be lowered slightly and made *a lot* more comfortable. To see if yours might qualify (before you buy), hold them up against the wall. Does the sole lie flat? You've got a winner! Easy adjustment! Also, look for shoes with high heels and a platform (in style as I write, anyway), which are far more comfy.

You need cute and comfortable shoes for your morning commute.

No one wants to trudge to the office in working-girl Reeboks over pantyhose (and really, pantyhose?), but limping twenty city blocks in a pair of stilettos is even worse. Luckily, there are lots of stylishly happy mediums. Stacked-heel, knee-high boots, and ballet flats are all chic options; add insoles to boost arch support. If you want added height, try walking-friendly wedges or platforms. Even hippie-chic clogs can work with tights and minis, and they're easy to toss off under your desk.

Your weight has changed and you no longer fit into your favorite dress, top, or jeans.

If you have lost weight, you might consider having your beloved piece tailored to fit your new body. Have your tailor leave generous seam allowances to accommodate any future weight fluctuations. If you've gained weight but don't want to miss out on wearing your lucky jeans, consult with your tailor to find out whether you can eke an inch or two out of them. If making them grow is a no-go, get rid of them. Hanging on to clothes that no longer fit won't do your self-esteem or your closet any favors. If the item in question is precious or unusual, consider putting it into storage for future generations of fashionistas.

So You Have to Buy a Bathing Suit

According to *Women's Health*, the average woman would rather undergo dental work than shop for a new swimsuit. Makes sense—both are painful but necessary. Here's how to make it out of the dressing room alive.

Play Make-Believe

To make the experience more enjoyable—and to have the most accurate idea of what you'll look like when you're actually wearing the suit—get ready to go swimsuit shopping the same way you would get ready for a day at the beach. Go to the gym, shower, shave, and douse yourself in body bronzer or tanning cream (the best choice for disguising flaws quickly). Just as you would if you planned to spend the day poolside, drink plenty of water and avoid sugary drinks and salty foods—a bloated belly does not a hot swimsuit model make.

Make Peace with Your Body Type

If you can't relate to Gisele, check out pinup photos from the 1940s, '50s, and '60s to see some real bodies looking really phenomenal and to discover ways to accentuate your most valuable assets. Before stepping into the horribly unflattering light of a department store dressing room, make an honest assessment of your body. What are your best features—your bodacious cleavage, killer abs, pert bottom? Which do you want to downplay—your full hips, small boobs, boyish waist? To flatter a bootylicious backside, look for a bikini bottom that ties on the side and hits low on the leg. Long legs? Look for swim bottoms with high-cut sides. To accentuate décolletage and deemphasize your stomach, try a fitted-bodice tankini. No boobs? Forget padding, it looks fake; instead look for a bandeau style top or a deep V-neck to add subtle dimension. Huge rack? Show off your curves with cups with side darts and seaming. To play them down and add support, look for wide straps and underwire styles.

Fit Your Lifestyle

Are you the spa-and-chaise kind of gal, or are you more likely to be found catching waves on your surfboard? If the latter, look for supportive styles that will save you embarrassing moments and provide you comfort and mobility when you're running around. Forgo the triangle-top string bikini and go for a stylish maillot or racerback tank.

The Devil Is in the Details

Approach brights and prints with caution, as they tend to accentuate flaws. In the same vein, look for small features such as ruffles, bows, belts, beading, and ties to subtly bring attention to the sexiest parts of your body.

Don't Underestimate the Add-ons

If bathing suit shopping has truly gotten you down, fret not. When all else fails, think like Kate Moss by adding personality to a basic black suit with cheap-chic costume jewelry. This will draw the eye toward your face instead of your tush. And don't forget, you can always be a wrap star—sarong, so right.

If All Else Fails . . .

Consider ordering several suits from J. Crew. They are fashionable, mixable, and size-forgiving. Order in multiple sizes and try on in the comfort and privacy of your own home. Send back what you don't want in a timely manner so your suits don't wind up costing you double!

In Shopping

It's amazing how clearly one remembers a bad—or good—shopping experience. For every bitchy salesgirl who ignores you until you come back to the store with Richard Gere, there's a kindly fashion guru waiting to make you feel like Cinderella.

When I was in college, I was dating a guy who had just graduated, and I went to visit him in the big city. When I got there, he told me he had forgotten to tell me we'd be attending a fancy party. I had nothing to wear and very little cash.

I went to Nordstrom, and the loveliest salesperson helped me find an amazing cocktail dress and pair of shoes, both on sale. I put them on and felt pretty amazing, but she looked me up and down and said, "Now what you need is a great pair of earrings." She led me to the jewelry department, where a cluster of major chandeliers were hanging like ripe berries on a vine. I wanted to pick them, but I was out of cash—and frustrated that my fairy godmother was now trying to upsell me when I'd already spent all I could.

"I love them, but I just can't afford one more thing," I told her.

"You need them," she said. "You need to look perfect tonight."

It took a lot for me to hold my ground, but I did: "Nope. No more money."

And she did, too.

"Here's what you do. Charge them to your Nordstrom card, wear them tonight, and bring them back tomorrow. I will personally make sure the return goes through without a hitch. Then one day, when you have a lot of money to spend on a wardrobe—and something just tells me you will—I want you to come back and spend it at Nordstrom."

Well, I wore those earrings, and I looked fab. And I returned them the next day. And a couple of years later, when my mom offered to buy me a work wardrobe for my first grown-up job in advertising, I traveled from our home in New York City all the way to New Jersey, just so I could buy everything at Nordstrom. I've been a loyal customer ever since.

That lovely salesperson turned what could've been a totally frustrating experience into a fond memory. Here's a few more ways to turn your shopping nightmares into daydreams.

A salesperson is blatantly rude to you.

If you're comfortable being direct, you can let the salesperson know that you think her behavior is rude. Start off calm and understanding. Lots of people are underpaid and poorly trained. Their incompetence might be the fault of management. If things escalate, be sure never to make threats you can't keep. Unless your family owns the company, you probably can't get the sales associate fired. And while it may feel good to vent in the heat of the moment, do you really want to be that girl?

If you'd rather avoid a face-to-face confrontation, get the salesperson's name and ask to speak to the manager in private. Remember not to seem accusatory—that's no way to win them over. Just explain what went down, and how you'd hate to stop shopping at their store just because of one nasty incident, but it's important the rudeness be remedied. You may get a discount out of this method.

If you're embarrassed and want to get the heck out of there, simply leave the store, craft a brilliantly cutting letter of dissatisfaction, and post it on Yelp.com.

You're undercharged.

The answer here is obvious—as tempted as you may be to pocket the extra cash, you need to tell the salesperson about the mistake. You may not pull an instant discount, but you won't spend the day feeling like a thief.

Your friend asks you what you really think about what she's trying on.

Use your judgment. Does your friend really want to know what you think, or does she want you to confirm her opinion? We have a tendency to squint and lie to ourselves when it comes to trying on clothes, so do her a favor and be honest enough to save her from looking ridiculous. But by no means should you tell her she looks terrible. Instead, suggest something else that would showcase her hourglass figure or bring out the green in her eyes. Something is wrong with the garment, never with her. And if your friend is supersensitive, you can always make it about budget: "Do you really think that fuchsia halter top is worth $600?" Chances are she'll get the picture without you having to hurt her feelings.

You're shopping with a friend and you both fall in love with the same bag.

First off, is there only one of the magical piece in question? Flip a coin to decide ownership, then agree that whoever loses the toss gets borrowing privileges. Or give in with grace and let your friend rock the bag—after all, a little good shopping karma never hurt anyone.

Are there two identical bags? If the tote is classic, you may each give it such a personal spin that onlookers won't notice you are carrying the same bag. On the other hand, if the tote is a distinctive, signature piece, you might consider splitting the cost and sharing the bag.

See above, except it's a dress.

Consider how different you look in it and whether you'd be likely to wear it at the same time. If you don't work together, it doesn't matter if you'll be rocking the same shirtdress to your morning meeting. But if you're each on the hunt for the perfect cocktail dress and you have the same group of friends, it's time for a compromise. Play a game of rock, paper, scissors for the big buy and require the graceful winner to lend out the frock for a less special occasion.

You could also both buy the dress and use it as an excuse to throw a twins party. Just as you shouldn't let commerce stand in the way of friendship, don't let individuality stand in the way of an awesome theme.

You've spent your last penny, but you need to look superchic (and loaded) at your high school friend's wedding this weekend (where all of your ex-classmates will be) and have nothing to wear.

Okay, we all want to look richer than we are (except for the truly rich, who seem to always try to look disheveled), but how do you pull it off looking like a knockout, not a knockoff?

- First rule of thumb: Go simple. The more flamboyant the look, the more likely it is to look cheap.

- If you have time and even just a few bucks, jump onto any one of the online sample sale sites (Gilt.com or Hautelook.com are my favorites). You'd be amazed at the discounts you can get on big designer names.

- Or consider renting anything from a designer handbag to a diamond watch at Avelle.com. That's right—you can actually *rent* a fancy bag, sunglasses, watch, or jewelry for a week or an event to make you look a lot fancier than you are.

- Not ready to lease? Consider fake diamond studs or a pearl necklace. For the studs, don't go bigger than the size of an eraser or no one will believe they are real. For the pearls, look for ones slightly large than pea-sized, not too shiny, and slightly off white.

Why Target Is My Favorite Therapy

Okay, I don't want the psychologists of America coming after me—therapy is fantastic for many. But for me, there is little an hour at Target can't solve. I consider it my own treasure hunt and claim victory over my wallet and my peace of mind each time I leave.

- **Fashion.** When it comes to trend pieces, I've given up fancy department stores. For anything that will be out in six months, go for the less expensive version. My chic aviators? $12.99. My python tote? $22.99. My leopard flats? $14.99. I think you're starting to get the picture. Each one has drawn compliments, and I never lie—I proudly claim my purchase place and price. After all, with all the money I save on trends that will change in six months, I can afford a really nice cashmere sweater that will last forever.
- **Entertaining.** I used to think I needed every color of china, every new place mat, every fancy vase, and so on. But

You bought something on final sale and realize that you hate it.

Final sale, schminal sale. First try talking to the store manager as soon as possible after you've made the purchase, to see if it's a firm policy. If you're a loyal customer, the store may prefer bending their rules to losing your patronage. Offer to accept store credit—even a restocking fee—if they'll accept the return.

If you can't sweet-talk the shop owner, consider offering the item to a friend as a trade or at a discounted price, or selling it on eBay. Always remember to try everything on before buying it, even if it means stripping down to your underwear in the middle of a frantic sample sale. Wear a loose dress over leggings for ultimate ease, and be merciless when you face the mirror. Never be seduced by a bargain-basement price; it doesn't matter how cheap something is if you won't wear it.

soon my pantry started to look like a garage sale reject, with mismatched everything. I never had enough of anything for the party I wanted to throw, and nothing ever seemed to coordinate. My new plan: buy the basics cheap. I have square white platters, plain white cake stands, simple white bowls. And I use flowers, candles, leaves—pretty much anything I feel like—to dress it up. Everything matches white, and having saved so much, I no longer feel bad when I want to splurge on a vintage candle holder or something fun.

- **Gifts.** I have two kids who between them get invited to a zillion birthday parties per year. While I tried for several years to be the essence of personalization, I gave that up when I decided to write another book. My solution: buy in bulk. Pick a great boy present (Nerf Dart Tag is a current crowd-pleaser) and a girl one (the make-your-own diaries and newfangled Shrinky Dinks have never failed me). Find one that spans two or three years in your age range, prewrap, and you're ready to go. Trust me, you'll thank me later.

- **You, you, you!** The little magnets that you needed to fix the door that wouldn't close, the cute makeup case that will be perfect for your weekend away . . . you can find them there if you look, and they're a bargain. Again, I'm not saying $150 for lying on a therapist's couch isn't worth it, but I've spent a lot less on Target retail therapy and felt great!

PART 7

Sticky Situations in Dollars and Not-So-Common Sense

You've probably heard that a woman's life expectancy, on average, exceeds a man's by more than a few years. It follows, then, that we should be even more concerned about saving for our futures than they are. Not so, though. According to Hewitt Associates, a research firm, women will need to save 2 percent more than men for thirty years in order to continue to live in the manner to which they're accustomed—a manner that, for many of us, includes too many pairs of Manolos and not enough mutual funds.

Even though, with the recent market downturn, we're spending less on fancy stuff, most of us still have no idea what it takes to balance our checkbooks—or our financial lives. I can't tell you how many women I know who earn sizable salaries but fritter them away on incidentals—spa pedicures, two-pound salad bar lunches—instead of saving them for big purchases. Gone are the days when women could depend on men to buy the house and make the car payments; studies show that most men today expect their partners to contribute equally to their joint financial security. And it's no longer cute or funny to be clueless about cash. It's not 1950, and you're not Marilyn Monroe.

So, where to begin? It's as simple as educating yourself. Start opening your bank and credit card statements. Open a savings account. Build a credit history so you can buy a home one day. Even though getting your finances in order can seem daunting, the freedom it brings is definitely worth the effort. And those new shoes will feel a lot more comfortable when they're not stuffed with guilt.

Your credit score is abysmal.

We're conditioned to approach credit with caution for good reason. Your credit score controls so many aspects of your life—whether you can rent an apartment, get a car loan, qualify for a credit card, lock in a mortgage. Poor credit increases what you pay for car insurance and may even affect your chance at scoring your dream job—more than 70 percent of employers check credit to judge your stability.

You have the right to a free copy of your credit report once a year, so without further ado go to AnnualCreditReport.com. Once you get a copy of your report, comb it carefully for any mistakes or incorrect details and then contact both the reporting credit agency as well as the creditor to notify them of any problems.

Once you know the damage, take action! Here are a few tips that will send your credit score into the stratosphere.

1. **Pay as much as you can.** The credit-scoring formulas like to see a nice, big gap between the amount of credit you're using and your available credit limits, so the best place to begin is by paying down your credit cards. Racking up a big balance, regardless of whether you pay your bill in full every month, makes a huge dent in your credit score. By simply limiting your charges to 30 percent or less of a card's limit, you can increase your score.

2. **Dust off an old card.** The older your credit history, the better. Experts recommend using your oldest card every few months.

3. **Ask for a break.** If you've been a good customer, a lender might agree to erase one late payment from your credit history. It never hurts to ask.

4. **Re-age.** A longer-term solution for more troubled accounts is to ask that they be "re-aged." If the account is still open, the lender might erase previous delinquencies if you make a series of on-time payments.

5. **Pay on time.** The irony of late or missed payments is that they hurt a good score more than a bad score. If you already have a string of negative items on your credit report, one more won't change much, but it's still something you want to avoid when trying to improve your credit.

6. **Resist the urge to get another card.** Applying for a new account can hurt your score, yet in general it is best to have smaller balances on a few cards than a big balance on one. Oh, the irony—how can you ever win?

Your identity gets stolen.

First, place fraud alerts on your credit reports. Close the accounts that you believe have been tampered with or opened fraudulently. In addition to speaking with someone in the security or fraud department of each company, you must make sure to follow up in writing, including copies of all supporting documentation. Next, you'll need to file a complaint with the Federal Trade Commission, and then bring a printed copy of your complaint form to your local police department so they can incorporate it into a police report. Be patient yet persistent in your follow-up with creditors and credit bureaus—you'll need to monitor all your accounts very closely for at least a year. When you begin to reestablish your accounts, consider using your full middle name, so you can easily identify them.

You need extra cash but don't want to be a stripper.

If getting another job is out of the question, open an eBay store! Sell those shoes and bags that have been gathering dust in your closet. If you've got kids, try selling their old stuff on HandMeDowns .com. Look for opportunities online by taking surveys (but do *not* agree to meet anyone in person). Call local market research facilities in your area and offer to participate in groups . . . you can make $50–$200 for an hour or two! Start a dog-walking business—you'll be able to quit your gym and earn some dough. If you're a bookworm, consider tutoring—post flyers in the vicinity of affluent schools and wait for parents to offer you $50 per hour to hang out with their teenagers.

You need to borrow money, but whenever you try, your tongue freezes up.

Think like a lender. What would make someone want to lend you money? The confidence that you'd pay back the loan. Regardless of whether you present your case to your parents, boyfriend, best friend, or bank, make them a reasonable proposition that's impossible to refuse.

Dear Jane,
I'm in between jobs right now and cannot afford to pay my student loans. What do I do?

Don't panic. Call your loan provider immediately! By law, every lender must offer programs for customers experiencing economic hardship. One option is something called deferment, which is a set period of time when you don't have to make payments and the government may pay the interest. If you don't qualify for a deferment, ask about forbearance, which lets you temporarily stop making payments and be considered current on your account. However, interest will continue to accrue, so you'll owe extra money at the end of your loan period. This interest snowball effect is no fun, so it's best to act fast. Don't think that avoiding the problem will make it go away. Showing that you can responsibly deal with it, even when you can't pay, is far better than avoiding a stack of unopened bills.

First, make an honest assessment of your life and finances. What income do you have coming in? What flows out? Present a well-thought-out list of dates and proposed payments—in amounts you can prove you'll be able to afford—and your chances of getting a loan are greatly increased. Second, think about the sum and why you are asking for it. If it's for that must-have new handbag, don't hold your breath. If it's for school, rent, or a new business, you stand a far better chance.

You've thought about getting a financial advisor but are worried you won't really like his advice.

I don't know about you, but when I was being told I should be a superwoman—have a job, give birth to three babies, cook a mean dinner, and service my man flawlessly—at no time did anyone give me any decent advice about money other than that I should want it. So when I actually do get it, I'm not sure what to do with it (besides buy nice shoes, that is). And regardless of your net worth or age, navigating the complex financial landscape is overwhelming. But it is best left to an educated professional, and you should at least get some advice even if you decide you don't want to take all of it.

There are a lot of people out there who call themselves financial advisors, consultants, planners, and other similar titles, and the varying sets of letters they use to express their credentials (CFA, CFP, ChFC, CPA, and so on) can be ridiculously overwhelming. So if word of mouth is not leading you to the right person, check out the Web site for the National Association of Securities Dealers (Investopedia.com) for a breakdown of what the acronyms mean.

You really want to buy a house, but don't know if you can afford it.

Like marriage, kids, taxes, and jury duty, buying your first house is a milestone of adulthood. As with any major purchase—car, computer, or Prada bag—the first step is to get your finances in order, to determine what you can spend.

The general rule is that you can afford to buy a house that runs about two and a half times your annual salary, and the down payment on a house is usually 20 percent of its price. As overwhelming and out of sight as this may seem, there are hundreds of calculators online that allow you to get a handle on your income, debts, and expenses and see how it all breaks down.

Don't forget that owning a house involves more expenses than renting an apartment does—you'll have to take care of the lawn maintenance, real estate taxes, homeowner's insurance, utilities, and gopher control. (Just making sure you're still paying attention.)

Obtain a copy of your credit report in order to clean up any errors, and make sure there will be no surprises when your lender runs a credit check. Another smart move is to get preapproved for a mortgage before you begin house hunting. This will save you the grief of looking at houses you can't afford and put you in a better position to make a serious offer when you do find the right place.

Now that you know what you can afford, do your research! Get to know the housing market backward and forward. Prices change constantly, and you have to do a lot of negotiation to get a fair deal, so make sure you do your homework before beginning your search. Another thing to consider is buying in a district with good schools even if you don't have children. You'll learn that strong school districts are a top priority for many homeowners, so buying a home in one will help boost the property's value when it comes time to sell.

Even though the Internet gives buyers unprecedented access to home listings, most new buyers (and many more experienced ones) are better off using a professional agent. Look for an exclusive buyer's agent, if possible, who will have your interests at heart and can help you with strategies during the bidding process. Like most things, word-of-mouth referrals are usually the best bet for finding someone you trust.

Once you've found your dream home, place an opening bid that makes sense in the context of similar homes in the neighborhood. Check out Zillow (see below) to get an estimate of value for the property and potential neighbors, and consider sales of similar homes in the past three months.

House-hunting Homework

Check out these sites before you begin to search to make the process less overwhelming:

Zillow.com has a reputation for delivering accurate estimates of value for properties. Check this out to get a good sense of what you can expect in your desired neighborhood.

NeighborhoodScout.com offers crime statistics as well as a breakdown of the types of crimes (violent or property) common to the area you want to live in.

If you have children and live in a state that participates in a sex offender registration program, it's not a bad idea to check out the relevant state's Web site (for California, for example, it's MegansLaw.ca.gov) to get an idea of registered sex offenders in your desired neighborhood.

Certain states have property legislation that requires owners to disclose to prospective buyers whether a house was used as a

Sticky As It Gets . . .

You're in trouble with the IRS.

Most of us manage to make it through tax season unscathed, but if you fall into the 1 percent of Americans facing audit, here's what you can do to make the process less of a headache. If you have an accountant, call him or her immediately (and start looking for someone else to do the job next year). If you file your taxes yourself, here are a few tips that help prepare you for the audit.

Take a good look at the details of your return and prepare yourself to support the items questioned by the IRS. If you feel that the matters being disputed by the IRS are over your head, it's to your advantage to hire a professional to step in and represent you. Before your audit, collect as much documentation as possible to prove the items/deductions in question, and take the time to organize and understand it before you meet with the auditor.

Experts agree that at the audit, it's best to speak only when spoken to and not to volunteer any information unless you're specifically asked. Leave your emotions at home and remain as professional as possible. Under no circumstances should you sign anything, especially Form 872, which extends the amount of time in which the IRS

methamphetamine lab. The federal government has recently established the first nationally centralized web resource for this, the Clandestine Laboratory Register. You can find it at usdoj.gov/dea/seizures/index.html.

can assess additional tax on your return. Finally, bring your checkbook and come ready to fork over serious cash if you can't support your deductions or if you have underreported your income.

Sticky Note: Some experts say that you are far more likely to get audited when you start earning less money after a period of earning a lot of money—i.e., the government wants to know why all of a sudden you're not banking so much dough. So while I wouldn't ever want to say you'd be in the clear, as long as you keep paying your dues and keep climbing that ladder, conventional wisdom says you should be okay.

You can't seem to save any money.

If you're serious about saving money, you need to take an honest look at the way you spend your hard-earned cash. Generally speaking, there are two types of costs that eat up your paycheck: unnecessary and unplanned. To get a handle on your spending habits, start keeping a money diary and writing down every purchase you make. At the end of the week, decide what you can quit buying, and stick with it. If you're trying to save, here are a few habits that add up and are easy to cut out: eating out (this includes expensive coffee drinks), unnecessary driving (gasoline costs), and not being mindful of your energy consumption (turn off lights when you leave the house).

In addition to lifestyle adjustments, reassess the places you keep your dough. For example, if you get two paychecks a month, take one paycheck and put it into a basic savings account, then put any extra cash into a money market mutual fund. Not only will your money be fundamentally secure, but you'll also have a better chance of surpassing inflation.

Sticky Note: If you want to save but just can't seem to get motivated, spend a few minutes playing with one of the many retirement calculators available online. Once you see what it takes to support yourself as a very old lady, you may think twice before indulging in those daily fro-yos.

You want to live luxuriously, but you're broke.

I hate to break it to you, sister, but having all the material things you want isn't always possible. Instead of fretting over the things you can't have, find creative ways to live on the cheap. Check out local Web sites for free cultural activities in your city; challenge yourself to make the best dinner you can for less than $10; get outside and become more active. Or try dressing up more when you go out. Sometimes a glamorous outfit can lead you to more glamorous activities; it doesn't always work, but you'll be ready if it does. Living a luxurious life doesn't mean having a massive credit limit on your American Express Black card.

A friend is always asking to borrow money from you but never pays you back.

Although you may have to get over some feelings of guilt, in order to avoid losing both your money and your friend, a simple no is best. Be honest and to the point, saying something like, "I know you need money, but I'm not in a financial situation where I can afford to lend anything to anyone. Sorry, I wish I could help." If *you're* that friend, stop asking. Very few people have the resources to keep giving and no one deserves continual handouts.

PART 8

Sticky Situations at Work

The odds of winning the multistate Powerball lottery are 1 in 195,249,054. So you'd better make yourself at home behind that desk.

"That's why they call it work," one of my friends says whenever anyone complains about what they do between the hours of nine and five. And when people complain about work, most often it's not about their job per se but about the people and situations that come with it. Work is the adult equivalent of the high school cafeteria—and we all know what a source of angst that was. There's the popular crowd (marketing people), the jocks (business development), the cheerleaders (human resources) . . . the list goes on and on. But, unlike in high school, you don't want to entrench yourself in a crowd in a professional environment, since the penalty, when something goes awry, is much stiffer than staying home from the dance.

You send an e-mail you immediately regret.

Train yourself to double-check the "to" field before you send your e-mails. Mistakenly hitting "reply all" is the number one cause of e-mail gaffes.

Some e-mail programs and providers, like Outlook and Gmail, offer an unsend/recall feature. Learn to use this feature! Sign up at DidTheyReadIt.com for an e-mail program that sends you a return receipt the minute someone opens your message. While this won't undo the damage, it'll at least let you know whether the receiver has read your missive yet.

Once you know the deed has been done, you can choose whether to acknowledge your mistake. Nine times out of ten, it's better to apologize than to pretend nothing happened. It's better to do damage control in person unless the topic is really embarrassing and inappropriate to address face-to-face. For example, if you expressed frustration at the way a coworker handled a project, you can approach that person and say something like, "Sometimes I get overly emotional when I'm passionate about a project, and I didn't mean to take my stress out on you." But if the regrettable message was personal—maybe you hit "reply all" by mistake and added a nasty comment like "Of course Sharon wants to ban jeans in the office; she looks terrible in them!"—it's better not to address it directly. Instead, leave a handwritten note—the shorter the sweeter—and a small token like a daisy or a pack of M&Ms.

You need to negotiate a raise.

Few things are more uncomfortable than talking about money. I've found the only way to make these sorts of conversations tolerable is to dial down the emotional volume and keep everything as quantitative as possible. This will help keep both you and your boss from flushing with embarrassment every five minutes. Then follow these simple rules.

First, do your research. Salary comparison sites like Salary.com and Payscale.com give you an idea of what you should be making, based on your position and geographical location. If your current salary is lower than the average quoted on these sites, you have significant leverage. Then take a look at your current job description and make a list of all the things you do that go beyond it. If you want more money, you have to prove you're doing more than your share of the work. Finally, if you don't get the answer you're hoping for, request specific feedback and a minireview in three months. Although few people are getting raises these days, given the current state of the economy, you should be rewarded for good performance. If your boss doesn't have the budget to increase your salary, ask for perks like extra vacation days or a plum assignment you've had your eye on. However, be careful not to threaten or grandstand. Don't tell your boss you're going to quit if you don't get a raise unless you really mean it. This can backfire if they're looking for an easy way to downsize.

Someone steals your idea.

Before you tell me you sent it to yourself in a sealed envelope, let us disprove all those intellectual property myths: Copyright law protects "works of authorship," like books and screenplays, and is applicable only when someone copies you verbatim. Patent law protects "useful and nonobvious" inventions and processes, but you have to register for a patent. Trademark law protects businesses when someone else tries to use their logo. (Notice there's no such thing as "But it was my idea!" law or "Why don't I ever get credit?" law.)

If you find you're not protected by any of these, and most people in a traditional workplace won't be, then assess the importance of the idea versus the relationship you have with the person who you feel betrayed you. First of all, does he know he stole it? Some people *genuinely* think they came up with something when they didn't. You may be able to stop the madness by simply telling the person how you feel or pointing out clear examples.

If you determine the idea is more important than maintaining the relationship, have a formal conversation with the concept-napper. To determine the tone of the conversation—that is, whether it should center on the relationship or the idea—

assess the relationship. Is the person your peer or your superior? Do you feel personally betrayed or just professionally betrayed? In general, it's best to keep the conversation unemotional, but on select occasions when the person you're frustrated with is a close friend, you're allowed to address your personal feelings of disappointment.

In the United States, business and intellectual property laws can vary from state to state. You'll need to research the appropriate copyright and/or intellectual property laws in your state, as well as U.S. Copyright Office law and policy for federal statutes, and/or get expert legal opinion. For the latter, check with the American Bar Association for recommendations or go online.

Some helpful Web sites are:

alllaw.com/topics/intellectual_property/

dir.yahoo.com/Government/law/intellectual_property/copyrights

copyright.gov/laws

publaw.com

You have a crush on your boss.

Having something slightly illicit spice up your workday is fun, right? Especially if the context is a manager-and-subordinate relationship, which heightens sexual tension while giving rise to all kinds of delicious fantasies about who's really on top.

While office romances, even with a superior, are not the faux pas they once were, every workplace has its own particular set of written, unwritten, spoken, and unspoken rules, many of which involve policies specifically forbidding fraternization or any other blurring of professional boundaries.

The bottom line is this: Don't date someone you work for. It's bound to end disastrously. But here you are, with a good-looking, intelligent, and (hopefully) ummarried boss. It seems so normal to be attracted. Not at all like a slippery slope. So, what do you do?

- Don't tell him. (But if you're in a committed relationship, do tell your husband or partner, adding that you have no intention of acting on your crush. That small act of confession will let some of the erotic air out of your reveries by removing the secret, clandestine element of your crush and returning it to an innocent, harmless footing.)

- Don't say or do sexual things around him.

- Don't spend late nights toiling away on a project for or with him. Go home at the end of the day. You can finish the assignment tomorrow.

- Stay focused on doing your job in your customary effective, professional way.

If your crush is not just sexual but based on genuine admiration, respect, and friendship, then redirect your feelings toward being a caring, loyal, and supportive colleague. Stand up for your boss. Listen to him when he needs an ear. Give him the kind of nurturing attention you would give to any relationship you value. And anytime you feel yourself wanting more, run for the hills.

You're sleeping with your boss.

Okay, so you missed the paragraph above? All righty. Not a good idea, even if you understand that this sort of relationship is a minefield through which you must proceed with extreme caution and discretion. Can't you just flirt with your boss and sleep with the water meter reader? Why do you have to play with fire? Because he's hot? Because your calculating mind believes it's the fastest route to a promotion and/or a pay raise? Because secrecy and being furtively felt up in a staff meeting are the biggest turn-ons you can think of? Because you're worried about losing your job, so a few rolls in the hay are good insurance? Because he's the love of your life and why shouldn't you meet The One here? (After all, most of us spend at least forty hours a week at our job site, which makes them social as well as professional environments.) Step out of your soap opera for a second and open your eyes to the reality of the situation.

Only you can assess your particular situation and circumstances, just as only you can define your self-worth, conscience, and moral values. However, whether your fling is an occasional, delicious, mutually agreed-to romp at the Ritz or an anchor-dropping affair of the heart that has you talking marriage and children, there are tips for handling the dynamite so it doesn't blow up in your face.

Don't kid yourself that your amour is a well-kept secret. Reality check: Everyone knows! And they're gossiping. Relationships between colleagues stir up feelings of anxiety and jealousy between even the most supportive colleagues. Rest assured, you will be talked about—and not kindly. Can you handle this type of negative attention? Your relationship and your character will take center stage.

If your strategy is to work your way up the corporate ladder on your back, get back on your feet and listen to this: In 2005, the California Supreme Court determined that sexual favoritism can, in some cases, create an abusive and hostile working environment that is an actionable statutory offense. Other states are following suit. Translation: Your trysts could be kinda illegal. Sexual jockeying to pull off coups at the expense of your equally, and legitimately, qualified coworkers being considered for a promotion could end up in a lawsuit that could ruin your career and your boss', too.

Accept the fact that no matter how the affair ends—in marriage or in flames—your job will never be the same. Things will be strained. Think about quietly distributing your resumé, seeking a transfer to another area of your company, or taking other steps to find new work.

Do not try to manipulate your boss or use sexual harassment laws in your favor. In many cases this backfires, leaving the subordinate jobless and blacklisted. Use the law only if you have been victimized, your mental stability is at stake, and/or you feel unsafe. If that's the case, seek outside counsel.

If this is a mutually enjoyable but not serious affair, you genuinely like him, and you want to be friends beyond the obvious hormonal attraction, play it cool. Push no agendas. Be discreet. Be his friend as long as he continues to be yours.

If you truly love him and want to give this relationship your best shot, let him know how you feel. And if he doesn't want you with equal ardor, be prepared to call off the affair and quit your job. Once you've slept with a boss you're in love with who doesn't love you back, your career path is bound to be short.

Your boss makes an unwelcome pass at you.

Continued, unwanted sexual advances or conduct on the job that creates a hostile, intimidating, or offensive working environment is considered sexual harassment, and sexual harassment is a form of sex discrimination that is specifically prohibited under Title VII of the Civil Rights Act of 1964. The first step in dealing with sexual harassment is to let the harasser know that the behavior is unwelcome. Confront your boss immediately and tell him his suggestions are unwelcome and unacceptable and you want them to stop at once. It's critical to communicate clearly and unambiguously. Office relationships often include a flirty element, and it's your responsibility to let your boss know his behavior was offensive to you, not funny.

If the harassment continues, or if you're uncomfortable about confronting him face-to-face, write a letter demanding that the behavior stop. Save a copy. In fact, save all communications—voice mails, e-mails, text messages, handwritten notes—that document your complaint. If the come-on was a voice conversation, keep a journal of exactly what was said—place, date, time—and if there were consequences to your denying him. Should you need to prove your case at some point, you'll want as much evidence as possible.

If the harassment still doesn't stop after you confront your boss, follow the sexual harassment or complaint policy in your employee handbook. If there's no handbook or policy in place, here's how to proceed:

1. Contact your boss' supervisor. You'll want him or her on your side if things escalate, and you don't want the supervisor to feel blindsided by a complaint to human resources.

2. With (or even without) the support of your boss' supervisor, head to your firm's HR de-

partment. Let them know how you've tried to handle it thus far and show them your documentation.

3. If for some reason you don't get satisfaction from inside your company, contact either the Department of Labor (dol.gov) or the U.S. Equal Employment Opportunity Commission (eeoc.gov). According to insiders, go to the Department of Labor first if you want to keep your job. The EEOC is a federal agency, and if they act on your complaint the company will come under federal investigation.

4. Consult an attorney. If you plan on taking legal action against the company, it's best to do this sooner rather than later.

Your boss says something offensive.

If the offending remark has to do with gender, sexual suggestiveness, or unwelcome comments about your body, it falls under the category of sexual harassment and you should treat it as such. If it's just ugly, puerile, and stupid, still confront him. Don't bristle and spend the rest of the day too angry to do your work. Tell him his remark bothered you and why. To keep it in the context of correcting the behavior and not matching it insult for insult, perhaps go to the restroom and count to ten, or get a cup of coffee, or close your eyes for a moment. Recognize that this person has a problem. It's not your problem. Your goal is to communicate that you find his remark distasteful and unacceptable and then go on about your business.

Your boss keeps passing you over for promotions.

Getting passed over for a promotion is painful because the rejection feels personal. When it happens repeatedly and you feel you were a solid candidate for each promotion, it's time to assess the situation. You need to know if you're on a track to advance within the company or if you're stuck. Losing out on a promotion you wanted, or even several promotions, isn't the end of the world. It's not the end of your career, either. Successful people use rejection as a learning experience. Don't let it immobilize you. Use it as an opportunity to develop better skills and to become more resilient and confident. The important thing is to get feedback and clarity. Then you can figure out what you need to do to change.

What's the first step? Dealing with your feelings. Take time to process your emotions, but do this with someone outside the organization who can help you reflect on your situation and figure out what you really want. Is your goal to expand your role? Change your job title? Get a pay raise?

Receive professional training? Reapply for the position you lost out on?

Then schedule a discussion with your boss. Your greatest need at that meeting will be to understand the why behind your failure to move up. To do this, you must stay nonconfrontational and open. It's information you're after, not the opportunity to argue your case. Don't beat around the bush. Admit that you're disappointed, but assure your boss you are not resentful. Focus on asking critical questions about what has happened and why. Has your performance been up to par? Are you as qualified as your competitors? What might you have done to improve your chances? What qualifications did you not have? Will there be other chances for promotions? What can you do to improve your chances the next time there's an opportunity to advance? Ask, and then listen carefully and attentively. Try to hear both what's being said and what's being implied between the lines.

Once you've gotten your boss' feedback, you'll have a realistic view of where you are. Now you can intelligently evaluate your circumstances and decide if you have a future in the company, if you can be happy there, or if you need to seek out a more rewarding situation elsewhere.

Remember, the important thing is to take the initiative. Irrespective of the outcome, finding out where you stand is an advancement all on its own. You'll learn from the experience and become more resilient, confident, and directed.

You're always covering for your boss.

Basically, if you work for an inept or incompetent boss, you have two options. One is to be loyal and camouflage his or her inadequacies. The other choice is to take advantage of the situation and expose the boss' flaws. There are pros and cons to doing each.

Do you genuinely like your boss? If he or she is great in some areas—hopefully the ones that matter most—and weak or inadequate in others, you'll take on more work but also be given the chance to learn more and gain valuable experience. Covering for your boss gives you the chance to shine and will also build your boss' trust in you. Plus, in most situations things tend to work out best for everyone overall if people work as a team, picking up slack for one another and working together toward a common goal. But never let your covering become lying to other employees or supervisors. You have an absolute right to your integrity.

Of course, you could also see this as an opportunity to usurp your boss' power and position and expose him or her to the right people at the right time, resulting in a wonderful promotion for you. This is a gamble that tends to work out better in novels and movies than in real life.

Whether you cover for the boss or not, the important things to cover are your own responsibilities, your mouth, yourself with a paper trail, and your own career. Do the job you were hired to do. If necessary, draw up a written job description for your position. Develop strategic partnerships and cultivate allies within the organization. Who knows—down the road these relationships might serve you well. At some point, a boss who's truly incapable will be exposed.

Your boss minimizes you in meetings.

In the corporate world it's critical to be considered a team player. So how do you get proper recognition for your contributions and be a team player at the same time? Continue to do outstanding work, and keep your boss' boss in the loop. Speak about your boss and his contribution to your work in a very positive manner. Even as you outshine him, make sure some of that glow is reflected on him to improve his status also. Don't try to take him out of the loop or blindside him. Meanwhile, in meetings, don't contradict or take issue with him but assert yourself in subtle ways. Be respectful and deferential, even when he's taking credit for your brilliant idea. But also make sure you know in advance what the agenda is going to be, then find a way to expand on each topic he's going to address. Whether it's zeroing in on a fine point, moving out into the bigger picture, or sharing facts or a piece of research you just happened to come up with, bolster and support his efforts while calling attention to your role and input.

Dear Jane,
I'm not a personal assistant, but my boss constantly asks me to pick up her dry cleaning. How can I ask for more stimulating tasks but still seem hardworking?

It's one thing to drop off an important letter or to pick up lunch for your boss if she's really busy. Asking you to run personal errands on a regular basis, however, is inappropriate, particularly if this wasn't spelled out in your job description. The best way to tackle the problem is to talk to your boss. Explain that you don't mind helping her out when she gets in a jam, but the time you're spending on her personal errands is keeping you from working as hard as you'd like and learning more. In fact, you'd been meaning to ask her to send meatier assignments your way that will allow you to really dig in and grow. Be sure to emphasize that doing her personal business every now and then is perfectly acceptable, that you enjoy being able to lend a hand. That way, there's a better chance she'll hear the part about wanting more stimulating work.

Your boss wants to be BFFs and you want to keep a safe distance.

The Best Friend Boss (let's call her a she because they usually are) is essentially a people pleaser who cares as much about being popular as doing her job. At first, she seems too good to be true because she's so interested in you and asks you to accompany her to ritzy business events, but after a while the constant questions about your boyfriend or your marriage, not to mention always trying to sort your life out for you, feel more claustrophobic than being taken under her protective wing. The best way to deal with her is to keep your distance—gently but assertively. Don't be available for after-hours socializing, and cut off personal chitchat on the job. When she asks how you're feeling, say, "Fine. I'm happy here and I appreciate that I can come to you whenever there's a problem. Right now, I'll just move forward on this report that has to be finished by the end of the day."

And don't automatically friend her on MySpace or Facebook, either, though she will no doubt friend you immediately.

You can't seem to make friends at the office.

Making friends at work can spell the difference between loving and hating a job. It can also change your work experience. Studies have found that people with one friend at work are statistically more likely to find their job interesting. People with three or more office buddies are more satisfied with their lives overall.

That said, social interaction on the job can be difficult for many reasons. Maybe you're shy about going up to people and saying hi. Maybe you're one of those people who has as many grumpy days as cheerful ones and sometimes it's just too much of a struggle to be perky and effervescent. Maybe you think of yourself as kind of weird and special—you'd just as soon spend your lunch hour doing push-ups in the park as sitting in the lunchroom talking about your kids.

The main thing to remember about communication is to always play to your strengths and look for ways around your weaknesses.

If you're afraid of rejection, then make people reach out to you. Bake cookies for the people in your department you have contact with. Send them an e-mail explaining how much you appreciate their help and that these cookies are your way of saying thanks.

Become a good empathizer. An ability to listen provides valuable on-the-job stress relief. People will soon know that they can come to you for support and a friendly ear. (Note: This is not the same thing as agreeing with everything they say or saying that you also share all of their gripes or troubles.)

FYI, former Massachusetts governor Jane Swift drew a $1,250 fine from the State Ethics Committee for having aides babysit her infant daughter when she was lieutenant governor.

In a 1995 survey by the International Association of Administrative Professionals, 54.7 percent of respondents acknowledged running personal errands for a supervisor and 67.1 percent said they had seen others do so.

Share your quirkiness with others. We all love people who accept themselves without judgment because it means they will accept us, too, flaws and all.

You have to fire a friend.

They say never to mix business with pleasure, and most of us know countless reasons why. Being forced to deliver some life-alteringly bad news to someone you adore is nothing short of torture. The best thing to do is just be straightforward; explain to your friend the reason you're letting her go, and offer to do your best to help her find her footing somewhere else. Don't sugarcoat—if she's being fired for poor performance, you'll be doing her a favor by explaining why so she doesn't make the same mistake again. If you were friends before working together, reassure her that your extracurricular relationship will remain unchanged. If you never should have hired her in the first place, admit it, but don't recount petty grievances (like the times she took off early to get ready for dates because, as her friend, "you'd understand"). Help her understand that while jobs come and go, good friendships are truly hard to find. If you became friends in the work environment, tell her that your fondness for her as a friend hasn't changed and that you'd love to maintain contact if she's comfortable doing so. And no matter what sort of future contact you agree upon, write a note thanking her for her work. It's surprising how little it takes to make the most awkward situation more humane.

If, on the other hand, you are the friend who got fired, take a moment before you send a company-wide e-mail featuring that picture of your boss/friend doing a bong hit back in high school. Is there merit to what she has to say? *Did* you take advantage of the boss/friend status? If so, chalk this up as a lesson learned and focus on getting another job and mending your bridges. If you feel this was truly without reason, think about whether your friend had no other choice—was she pressured from above? Talk to her as a boss, not a friend. But try not to put anything vicious in writing. Trust me, you'll regret it later.

A colleague overhears you talking badly about him or her.

Number one, stop talking. Your opinions about coworkers, even if they are work-related rather than personal, fall under the umbrella of gossip. As tempting—and human—as it can be to wonder to X why Y is putting on all that weight, or why Z can never get her assignments done on time or slips out early, your best chances for survival and advancement depend on staying deaf, dumb, and clueless when it comes to the office pipeline. What goes around comes around—if you talk about others, others will talk about you. If someone overhears you talking about him, woman up and go with the straight-on approach. First apologize, then eat crow: Promise that next time you have questions or comments about him, you'll come to him directly.

A colleague always takes credit for your hard work.

If a colleague tries to undermine your efforts and take credit, by all means make sure that your work and contributions get noticed. However, you also need to avoid coming off as too aggressive or defensive. Each situation will vary according to the specifics, but here are some tips for a measured approach that will bring you the best results.

Avoid dealing with the problem in e-mail. Personal conversations are the best route to opening lines of communication that will resolve, rather than escalate, the issue.

Keep your professional reputation intact by staying cool and composed. It may sound contrived, but writing a little sound bite for yourself that clearly states the problem without placing blame can help you stay focused and tactful.

If the colleague still refuses to acknowledge your efforts, you have to decide whether to take the issue to the next level by going to the boss with concrete examples of your input.

Finally, make sure the situation doesn't repeat itself. In the future, don't disclose ideas one-on-one with coworkers. Present them in team meetings or send them around via group e-mails. Protect yourself through documentation and public record.

You fall in love with your coworker.

The good news: Although the common belief is that office dating is a recipe for disaster, according to recent studies cited in *Fortune* and *GQ*, somewhere between 22 and 50 percent of office romances actually lead to marriage.

That said, for every office romance that pays off in gold and diamonds, there are many more that blow up like water balloons, dousing everyone involved. If you're determined to dive into the office pool for love:

- Take your time before hooking up, especially if you're new on the job. Establish yourself and let people get to know and respect you for the work you produce before letting your sexy genie out of her bottle.

- If you do have something going, don't talk about it—to anyone. Your long lunch might speak volumes, but don't let the first detail slip past your smeared lipstick.

- Maintain a strict line between the personal and the professional. Off-site, feel free to play the vixen or the angel, but once you punch the time clock you're colleagues.

- If you end up dumping him, continue to be cordial and collegial. If he dumps you, forgo the temptation to hang a voodoo doll stuck with pins and festooned with a lock of his hair or favorite tie on your office door. Remind yourself that you always knew your love entailed risk.

Someone finds out how much money you make.

A coworker coming at you with knowledge of your salary is like a coworker coming at you with a loaded gun. That information is like ammunition, and you should back away from it. Neither confirm nor deny; just say something like "Don't take it personally, but the truth is, I consider my salary to be as personal and private as my sex life and, frankly, I'd be more comfortable discussing the latter than the former." Then smile and change the subject.

Someone who you think does a less-than-amazing job asks you for a recommendation.

You can dislike someone and still respect him and the way he does his job. The first thing is to ask yourself if you and this person just don't click personally or if your antipathy extends to his professional skills as well. If the answer is yes to the latter, then just say no with the excuse that you are too busy or don't know him or his work well enough.

Even if you can vouch for his work, before you say yes be sure you can put your emotions aside and write a recommendation that is convincing and sincerely positive. A lukewarm, stilted endorsement that wraps its affirmatives in veiled criticism would be unethical and irresponsible. If you do say yes, remember that a letter of recommendation lives or dies on its concrete examples and evidence of someone's credible accomplishments. Most potential employers evaluate a candidate on general qualities like initiative, trainability, enthusiasm, leadership, self-motivation, adaptability, imagination, and communication skills.

You work closely with an acquaintance at your job and suddenly you're unwilling BFFs.

BFF-ship is a two-way street, not one woman unloading and confiding and bitching while the other holds up a garlic clove to ward her off. If *unwilling* is the key word here, then it's up to you to set boundaries and keep her at arm's length. How to deal: Be friendly but coolly detached. If she corners you with personal chitchat, don't engage by being too interested, or overflowing with sympathy, or full of advice. Just nod and respond with a very big-picture, impersonal perspective about how everyone seems to be going through something or

Dear Jane,
At the office, if I put my lunch in the shared fridge, what do I do when I encounter some slob eating it?

Be spontaneous! Do whatever occurs to you to put a stop such a heinous act! Make a citizen's arrest. Stab the pilfered sandwich with scissors. Grab your staple gun and his tie (if the pilferer is male) and nail him to the table. Say, "Hey, do you know whose short ribs those are?" or "Why are you eating my leftover pasta Bolognese?" Or say nothing at the moment, but later tack a note to the refrigerator door, something along the lines of "I hope whoever stole my sushi from the fridge enjoyed it, since I was going to take it back to [insert name of restaurant] after discovering a worm." In the future, you might draw a skull and crossbones in black marker on your lunch container. Or simply tag it with your name on a Post-it.

Mercury is retrograde. If she wants to get together after work, be unavailable. Every time she gets personal, turn the conversation back to the work at hand. Make work the rich and fertile ground between you where much is shared and much accomplished. And then let yourself off the hook about it not being a close friendship. Women, much more than men, tend to believe that if someone likes us, we have to reciprocate, or that we should be able to be friends with anyone. The truth is, sometimes the energy just isn't there and that's okay. There's no shame in being civil and cooperative and leaving it at that.

You have a genius employee who is a total slacker.

It's a conundrum, isn't it? Psychologists say we often do our best thinking when we're not concentrating on work at all. And yet, our basic work ethic tells us that nothing worthwhile is achieved without palpable effort, and that results trump talent—the employee who bothers to format the spreadsheet so it's readable is more valuable than the one who comes up with brilliant ideas while he's playing video games for an extra half hour after his lunch break.

Slackers are annoying because it's as if they're always calculating how to meet the basic requirements of a task or assignment with the minimum of effort and see no reward in exceeding that standard. Which drives us crazy—number one, because we know they're capable of so much more, and number two, because they're always forcing us to draw a line between slacker genius and plain old laziness.

If you're blessed, or cursed, with an exceptional, creative employee who believes procrastination is the mother of invention, you have several choices: You can clearly lay out what the tasks are and your expectations that he fulfill them, and then you can manage his time. You can say nothing and continue to seethe about lost productivity. You can remind him that while being a slacker genius is a social asset among schoolkids, working hard is a sign of being a grown-up, and someone who has dazzling ideas hourly yet fails to put any into practice is so not cool. Or you could consider what managers at Google headquarters in California have been doing for nearly a decade—encouraging their best and brightest to put aside assigned projects for a percentage of every day and pursue their own unique creative schemes. Who knows what allowing your slacker genius to stare out the window or otherwise noodle around might come up with? For Google it has so far produced Gmail and Google News.

No one can do things as well as you.

Okay, so you're an anal personality and a micromanager to boot. You're picky and controlling. You have trouble delegating. You're so obsessed with the details of every project you ignore your big-picture, long-term responsibilities; when you're out of the office you call in hourly, even if you're home with the flu. If someone shows up thirty seconds late for a meeting you need Thorazine. News flash: Your hovering presence doesn't ensure that work is done more creatively, effectively, or efficiently. In fact, more often than not, wielding the reins too tightly disempowers your workers. If you're constantly looking over their shoulder, it wastes time, kills morale, leads to quicker turnover, and keeps you from looking into the future and setting goals.

Why do managers become micromanagers? Sometimes they fear that talented subordinates might bypass them professionally. Sometimes pressure from senior management for better performance evolves into a compulsion to control the work. Entrepreneurs who are driven and passionate about their companies often find it difficult to let employees utilize their skills, abilities, and knowledge.

Here's an action plan to help you back off:

- Hire the best staff you can. Look for people who bring significant skills to the table and can apply them without direct supervision.

- Set goals and hold people accountable.

- Remember that in business, time is money. If you oversee every job in the office, store, or factory in minute detail, you're not valuing or maximizing the use of your own time.

- Provide the necessary resources for achieving goals. That could be software, professional assistance, additional training, or education. It's up to you to outfit your staff with the tools they need to get the job done.

- Focus on outcomes. Yes, you want the job done right, but right doesn't necessarily mean your way. Let employees take their own approach. If they're achieving the results you want, acknowledge that there might be a better way or at least another way that is as effective as yours.

- Do stay available. Just because you're no longer monitoring your staff's every move doesn't mean you abdicate your authority. When someone asks you a question, drop what you're doing and support them immediately.

- Listen to your employees. If they have an idea, hear it out. Brainstorm with them.

- Praise people. Acknowledging a job well done can become as habitual as micromanaging.

You're intimidated by the people you manage.

Here's what not to do: smack their heads from the back as you walk by, show them your middle digit from the front, or say outrageous things so they'll think you're weird and keep their distance. Instead, ask yourself why you're intimidated. Do this with paper and pen and write down the answers. Is it because they're older than you? Have more experience, which you fear will one day unmask you as being incompetent or not yet ready to handle authority? Do they hold advanced degrees from prestigious universities? Tend to be loud and pushy where you are subtle and reserved? Are you a petite woman supervising guys who look like a wrestling team or construction crew? Do you come from an immigrant family who arrived in America only recently, whereas your staff are Mayflower descendents? Whatever the reasons, getting your feelings down on paper is the first step toward demystifying their power.

Intimidation, you see, isn't about them; it's about the feelings of fear, awe, or inadequacy that others, for whatever reason, arouse in you. Your objective should be to replace those feelings with belief in your own abilities and the assurance to stand your ground. How to accomplish this? Begin by reminding yourself that your employers hired you for or promoted you to a leadership position. Clearly they have faith in you.

You might also think about whether letting

others intimidate you has anything to do with your need to be nice. This is a common problem for women in managerial positions. The assertiveness and power that go with being a boss sometimes collide with the messages we got as young girls. Girls are far more likely than boys to fear being disliked, or to worry that being in charge will render them unfeminine. Remember that giving up the need to be nice doesn't mean suddenly morphing into a dominatrix. It isn't about external behavior. Real authority always comes from within, and if you have it, you can speak as softly as you like, because your big stick is your self-confidence.

You're younger than your employees.

It's hard to direct people who are the age of your parents. And understandably, people may be skeptical of someone their daughter's age telling them what to do. Today's workplaces often have two or three generations working together, and while differences in work styles, work ethics, technical knowledge, and experience might cause some friction, the situation can function well for everybody. Here's how:

- Acknowledge their expertise and experience. Give them credit for the knowledge they have. Don't be afraid to learn from them. Try saying, "Tom, what do you think? You have lots of experience in this area."

- Share praise and give them access to upper management. Some of them may have wanted your job or feel they were bypassed by a young woman. If you share credit and give them exposure to those above them, they won't feel cut off.

- Don't micromanage them. Nothing irritates a seasoned professional faster than someone younger not giving him enough room to complete the job in his own way, as long as he reaches the desired outcome.

- Don't second-guess yourself. Remember that if one of them were qualified to lead this group, he or she would have been selected. Instead, you were chosen. Don't let anyone walk all over you. Insist on respectful treatment as the leader.

- Keep them challenged. The best way to keep employees motivated is with challenging work that suits their strengths and allows them to contribute.

- Accept that there will be differences. According to *Forbes*, older workers tend to believe in face time at the office. They show up early, work through lunch, and come in on the weekends. Gen Xers and Yers were raised in the Internet era, where it doesn't necessarily matter where work gets done, as long as it does. Technology can also be a separation point. Younger workers are more intuitive about computers and software applications, while older employees who grew up before Google, Facebook, MySpace, and Twitter might take longer to adopt new technology. But these distinctions can be positive.

You overhear an employee saying something unkind about you or another coworker.

Facing down gossip isn't rocket science, and hopefully you won't go ballistic in the process. In fact, it's pretty simple. Confront the people who are talking and tell them (without yelling) that you heard what they said and that you don't appreciate it. Tell them they're free to come to you if they're curious about what's going on in your life. Take the same tack if it's about someone else. Say something like, "If you want to talk about Rosalind, let's go get her so she can be part of the conversation," or "Hey, I'm really not comfortable listening to you two talk about Boris while he's in the next room." Don't lecture and don't berate. Just say your piece and move on.

You're interviewing for a job and want to know about, ahem, the bottom line.

The general rule is not to ask about salary, benefits, sick pay, or maternity packages in an interview. Asking for a particular figure could give the impression that salary is your major consideration in applying for the job. Don't even include salary expectations in your resumé. You should have checked that the job was within your target salary range before the interview by finding out the average salary for that position on the job market and then evaluating your own experience, expertise, and educational qualifications. Web sites that deal with employment and job opportunities are the places to do your homework. A friend of mine once interviewed a woman who asked nothing about the position except when the educational benefits kicked in and how big her office would be. Next!

Leave salary negotiations until the job offer comes through. Your negotiation position will be stronger then. Even if the interviewer asks, "What's your salary range?" don't name a number. If you give a figure higher than the range for the job, the interviewer will tell you you're high, and you've just lost money. If you give a figure lower than the range, the interviewer will say nothing, and you've just lost money. Ask the interviewer to tell you the range for the position. Then you can focus on getting to the high end of it.

PART 9

Sticky Situations on the Go

Once, on a flight from New York to San Francisco, I was seated with a fashionably dressed young woman about my own age, and we immediately hit it off. As we discussed boyfriends and handbags over bubbly water and warm nuts, her eyes suddenly rolled back in her head. Her top lip started twitching, and then her hands curled in toward her body. She was having a seizure.

I called for the flight attendant, but by the time she arrived, the woman appeared simply to be asleep. The flight attendant poked her gently to make sure she was okay, and said, "Your seatmate here says you just had a seizure. Are you feeling okay?"

The woman looked at me incredulously, as though I were making it up. "A seizure?" she said, "Don't be ridiculous. I'm fine! Look at me!"

And she did look fine. She appeared to have absolutely no memory of the incident at all. Now the stewardess looked at me as though I were crazy. I glanced around frantically for someone to substantiate my story, but the man across the aisle was looking away.

"I promise you," I said to the woman, who I knew needed to see a doctor ASAP, "you had a seizure. I've seen them before. I know what they look like."

"I find this whole thing very insulting," my seatmate said to the flight attendant, "not to mention bizarre. I'd like to move to another seat."

But there was no other seat, so I had to endure the remaining four hours of the flight trying not to melt under the seizer's hateful gaze. I'll admit a small part of me wanted her to have another episode, just so someone else would see it.

She didn't, of course, and she continued to glare at me all the way to the baggage claim. At the taxi stand, though, I ran into the flight attendant, who said, "Don't worry. I saw her have the seizure. But you know—the customer is always right."

So many bizarre things can happen when we're moving from one place to the next, watching strangers who quickly become intimates pass in and out of our lives. This chapter will help you make the transition as smoothly as possible.

On the Plane

A solid hour into your trip on a red-eye, your seatmate is still yammering like it's noon, not midnight. You need to get some sleep in order to function in the morning. How can you shut her up without making her uncomfortable?

I can't tell you how many times I've fought to keep my eyes open while nodding appreciatively as the little old lady in the seat next to me tells me about her new grandchild—for the fourth time. I'm not sure why I always find the need to make strangers like me, but try as I may, I just can't be gruff—only superpolite.

After decades of flying, though, I finally have an easy way to get the little old lady to zip it without worrying about hurting her feelings. I just say, "Please excuse me, I'm so sorry, but I just took an allergy medication that completely knocks me out, and so I must close my eyes." (I like to say allergy medication so they don't think I'm some refugee from *The Valley of the Dolls*. Here I am, caring what strangers think again!)

You're flying coach to a third-world country. You know it's going to be uncomfortable, but the experience is worth it, right? You find your seat and remember they don't really shower in said third-world country (the guy in the seat next to you stinks). Is this a cultural custom you can live with for eight hours?

Sorry, but yes, you're going to have to suck it up—or sniff it up, as the case may be. Unfortunately, once the seat belts are fastened, the overhead compartments are shut, and the door is closed, there's not much you can do about your seat assignment. You can always ask a middle-row person if he wants your window or aisle seat, but there's a good chance you're going to run into the same smelly situation—okay, twice the chance, actually—once you're sandwiched between two fellow travelers. I always travel with all-natural antibacterial wipes scented with peppermint, lemon, or lavender essential oils. You can use one to wipe your sticky arm rests, then conveniently hold it in your hand, near your nose, and breathe through it like a gas mask. Take a lot of walks and make friends with the flight attendant in the back of the plane. Maybe if you explain your dilemma, she'll let you sit in her jump seat for a while.

A woman you work with travels to Europe for business. She always wants to borrow an Ambien for her flight. You want to tell her to get her own meds, but what's one tiny little tablet?

I think it's actually a felony to share your prescription medication with others. It's just not a good idea to share pills, especially with someone you work with, and extra especially if she works for you. What kind of message are you sending an employee when you willingly supply her with the means to zone out on the job? And since no one really knows what people do while taking Ambien anyway—we've all heard about the sleepwalking, sleep-driving, and, worst of all, sleep-eating that occur while under the influence—it's probably best to take responsibility for no one's behavior but your own. A friend of mine was taking the red-eye from LA to New York. Before takeoff he was chatting, not flirting, with another guy in his row. He took his double Ambien and woke up a couple of hours later making out with that very guy. Who was married, by the way. True story. You just don't want to be that person—or, even worse, to be responsible for that person.

Jet Lag: Away-from-Home Remedies

Sometimes a simple flight can feel like an ordeal in time travel. When you cross more than one time zone, your body's systems can become confused and make you feel at once sleepless and exhausted, disoriented and edgy.

I've lost more than one European vacation to jet lag. One time when I went to Madrid with a friend on a trip we'd been planning for months, I slept all day and stayed up all night reading in the bathtub. I might as well have been in Peoria.

Thankfully, though, I've now discovered some remedies for jet lag that really work.

Planning Your Travel

If you're traveling east, plan a flight that drops you at your destination early in the morning.

If you're traveling west, book a flight that has you arrive in the late afternoon.

Don't schedule any important meetings or events for at least twenty-four hours after you arrive. This will cut down on your anxiety about being able to fall asleep.

On the Plane

Reset your watch to your new time zone the minute you settle into your seat. Behave appropriately for the time of day at your destination—if it's daytime, stay awake; if it's nighttime, try to sleep.

If you can stand it, skip the food on a long flight until the last meal, which will be served at the appropriate time for your new time zone. This will help your body adjust quickly.

Avoid alcohol and caffeine—artificial depressants and stimulants will only make things worse. Drink lots of water and herbal tea.

Some people swear by taking a dose of melatonin, a natural sleep-enhancing supplement, before bed in their new time zone. While I prefer chamomile tea, I have many friends who consider it a lifesaver. Consult your doctor if you're considering this.

Once You Land

Get some sun as soon as you can. Taking a brisk walk outside does wonders, and the sun resets your circadian rhythms.

Eat fresh fruits and vegetables. Their high water content fights dehydration, and their fiber encourages, ahem, your pipes to keep moving.

Don't beat yourself up if you're up all night. Don't waste your trip reading in the bathtub, like I did—wherever you are, chances are there's something interesting going on, especially at three in the morning. But be safe!

At the Hotel

You're on a solo vacation, relaxing, recuperating, recharging. You step into the Jacuzzi while waiting for your massage at your fancy hotel. A cute guy joins you and starts asking all sorts of questions: Why are you alone? Where do you eat dinner? Where are you from? Then he asks if you're ready for your massage.

Do not go into that massage room with him. Take it from someone who has survived this not just sticky but icky situation. There's not a chance in hell you'll enjoy this massage while wondering whether your therapist was coming on to you, and whether that article in *Allure* about happy-endings massages was for real.

Never take your clothes off while feeling vulnerable. Period. You're paying top dollar to relax not just your back but also your psyche. Whether you find the therapist attractive, creepy, or downright threatening doesn't matter—you'll be focusing on him instead of yourself.

Against my best instincts, I followed that therapist into the treatment room. Once I was on the table, he told me I was holding a lot of sexual tension in my lower back. *That's not sexual energy, it's self-defense!* I wanted to say. But I was dumbfounded.

Any aboveboard massage therapist will tell you not to go into a treatment room feeling uncomfortable, and that being totally honest is the only way to get a truly satisfying massage. But if you don't feel comfortable speaking to the therapist face-to-face, scurry back to the front desk and explain the situation. If you don't want to be specific, just say the therapist made you uncomfortable and you'd like to be reassigned. Or make up an excuse to get out of there ASAP—even if you must fib and say it's your time of the month.

Let's face it, your hotel room sucks. But you're traveling with business colleagues and you don't want to be the prima donna—again. Should you make a fuss or keep staring at the generator on the Astroturf outside your window?

Wave goodbye to your hallmate/business colleague, get into your room, close the door behind you, then call the front desk. There's absolutely no point in staying in a room that makes you feel uncomfortable when you're paying good money for it. Or at least your company is.

Dear Jane,

I recently went on a girls' weekend with my best friend from childhood. We hadn't shared a room since seventh grade, but I figured that since we talk on the phone all the time, how bad could it be? From the moment we got in the rental car, she was singing and telling dumb jokes—I couldn't get a moment of silence all weekend. Far from relaxing, the trip was exhausting—and more than once, I really wanted to kill her. Now she's asking me to be her guest at a luxurious wedding abroad. How can I say no without hurting her feelings?

This is one of those rare times when honesty isn't necessarily the best policy. Since you and your friend rarely spend time together in person, don't make her self-conscious by explaining that you don't want to go away with her simply because she annoys the hell out of you. Instead, fabricate a scheduling conflict and tell her you can't wait to see photos.

You're traveling with a friend to France. Like you, she's chic enough to carry off skinny jeans and

You're traveling to Brazil on a pretty rugged hiking trip. You've edited each and every item in your suitcase, bought wash-and-dry underwear, and tossed out the extra flip-flops—you'll go barefoot on the hotel carpet in an effort to be as mobile as possible. Your friend, however, shows up with a suitcase she can't even wheel around, never mind run with to catch a plane if need be. She's cramping your style, but you're already at the airport. Now what?

Pray for weight restrictions—on the luggage, of course. Then the airline will be the one to tell your travel buddy that she needs to pare down or else, and you won't be stuck pulling her weight. Help her edit on the airport terminal floor, and try not to make her feel bad for behaving like Paris Hilton in *The Simple Life*—after all, this is the beginning of your vacation. Escort her to customer service and ask them to ship her extra stuff back to her home address. These days, most people will do anything for a $50 tip. If they refuse to help, there's no need to throw all those jeans and boots away; most large airports have Federal Express kiosks with staff who will be happy to ship her stuff.

The next time you travel with a friend, discuss what you're taking before you get to the airport. E-mail a packing list to each other in order to avoid duplicates—and surprises. You can share toothpaste for a week. This will help you avoid the worst-case scenario, where one friend checks a bag and the other carries on. Now that's cause for a falling-out.

a good scarf. But on your first sightseeing day, she's dressed in camper shorts and sneakers. Who is this person? You are embarrassed to admit this—but you're embarrassed to be seen with her.

Suggest that your first stop be a shopping destination: "For fun, let's dress like we belong here." A beret and striped shirt is not as bad as a T-shirt and fanny pack (do they even still make those?). Plus, that Marseilles look can go with your skinny jeans and flats, and voilà, part *Teen Vogue*, part Musée d'Orsay. And hey, you might just fool a few tourists as well. Oh, and it's no longer funny to try to speak French when the ugly Americans fall for your outfit. Be nice and give them real directions in English.

Out to Eat

You need an emergency reservation.

In restaurants, it's all about who you know. If you need to get into a hot spot at the last minute, explore every possible connection you might have to the place. Whether you know the hostess from yoga class or the owner used to play golf with your dad, remind the restaurant of this connection when you call. Don't, however, manufacture a dishonest "in"—this strategy is transparent and will definitely backfire.

If you're just not a VIP, the best thing to do is be honest. Tell the person who answers the phone that you know this is crazy, but is there any chance there's been a cancellation? You just love the food and someone is coming from out of town and you'd positively die to have the opportunity to bring him or her in. You appreciate any help at all they can offer—and you'll certainly accept a place on the wait list (just call an hour later and sweetly ask to check the status of your reservation).

Of course, there are folks who swear that being pushy and aggressive is the best way to land a res, but I couldn't disagree more. Would you want to reward someone who was rude to you? In fact, if a kindly host or hostess does manage to snag you a hot table, bring a token of appreciation when you show up for dinner. A tiny box of good chocolates,

Dear Jane,

Every time I go out to dinner with my new boyfriend, he announces, as we walk toward the banquette, that he wants to sit "on the inside." Am I overreacting if I think this is not okay?

My response to this one may surprise you. Maybe you think I'll say, "Why not let the guy sit on the soft-yet-firm, cozy and comfy, best-view-in-the-house seat for a change? You're equal partners in the relationship, and he's going to pay for dinner anyway, right?" Wrong. I mean, he is probably going to pay for dinner—but the lady always sits on the inside. First of all, there's likely a mirror behind the banquette, which means the person in the chair will see him- or herself chew, swallow, and sip the whole meal. Girls hate watching themselves eat and become jumpy and self-conscious at the prospect. Guys don't notice. Second, the girl should a bottle of champagne, or even just a handwritten note will show your appreciation, and is much more chic than a wadded-up $20 bill.

I have one fabulous friend who always calls pretending to be her own assistant. She starts by saying, "Hello, this is Ms. X's assistant. Listen, we do *not* want happening tonight what happened last time. She's coming in with four people at eight and will need a very good table." She says it strongly and firmly enough that she has never been denied. I myself can't get past hello. . . .

always be made to feel comfortable at the beginning of the evening, since later on the guy is going to want to seduce her. Fair is fair. Remind him of this and he'll gladly surrender the seat cushion.

The woman seated next to you in a quiet cafe where you're trying to get some work done is talking on her cell phone—loudly. You can't concentrate. Should you say something?

Try to make eye contact with her, then turn your phone on and off in as obvious a manner as possible. But never be openly rude—you never know when that person will turn out to be the estranged sister of an ex, or a participant in a very small elite travel group of which you're soon to be a member. I once ended up sharing a tent in Tanzania with a loud-talking, cell-phone-loving lady I had met in a cafe a few months prior, and, you know what? That voice of hers came in handy when it was time to ward off the wild animals. A good rule of thumb is to always behave less obnoxiously than the person next to you.

You need to order pickily, but you don't want the waiter to hate you.

It's a big foodie trend these days to print a disclaimer at the bottom of your menu. "No substitutions—ever," declares a twenty-two-year-old chef who's just a teensy bit in love with his own talent of combining foods in bizarre and pretentious ways.

While I understand that cuisine is art, I believe that if you're going out to spend your hard-earned cash on a meal, you deserve to get what you want. With that in mind, there's a nice way to ask for it, and a disrespectful one.

Instead of choosing a specific dish and requesting modifications to it, ask your server if he can recommend something that doesn't contain whatever items you want to avoid. Give him the chance to show off the restaurant's menu—he may come up with an idea you love. This method also encourages—rather than limits—the chef's creativity.

If you want to order something that comes with a sauce you hate, ask for the sauce on the side, rather than not at all. Then just don't eat it.

If you have food allergies, declare them the minute your tush hits that banquette. No matter how passionate a restaurant is about its culinary integrity, no chef wants to have a patron die on his watch. Be firm about your dietary restrictions, and state—even overstate—the consequences ("I will die if there is even half a sliver of one pine nut in that pasta").

The Tipping Point

Here are the rules your server expects you to know. Use these as guidelines and go up a bit or down a bit, depending on the service you receive. And remember, no matter what the owner of a restaurant or nail shop says, tipping is always optional.

The amount you tip should always be based on the total bill before taxes. There's often a lot of confusion about whether to include alcohol in the total on which to base your gratuity. Use your discretion—if you drink one bottle of wine, sure, include it. But if the meal is preceded by a two-hour cocktail hour tallied on the same bill, tip on the food total only.

You should leave your server a tip equaling between 15 and 20 percent of the total bill if the service ranges between pretty good and excellent. You're sending a message that the service you received seriously sucked if you leave a tip of less than 15 percent.

Parties of six or more (I always thought that number seemed pretty arbitrary, but it's consistent) will usually

Ordering neurotically is one time it doesn't hurt to be self-deprecating. Say what the waiter is thinking before he does. "I'm sorry to be such a pain," or "I must seem like one of those crazy, food-phobic socialites" is a funny, lighthearted way to begin the process of getting what you want.

You told your date you love wine—and now he thinks you know how to order it.

I am of the opinion that you don't necessarily need to be a wine connoisseur in order to be a wine lover—if you know what you like, you know what you like. Still, there's an art and a science when it comes to ordering wine, and it helps to know a few basic rules.

A glass or two of bubbly makes an elegant way to begin a meal (and a fun way to end it). Just remember that the only kind of sparkling wine that can truly be called Champagne is from the Champagne region of France. Prosecco is the Italian version, and there are lots of excellent (and affordable) options that come from California.

The most basic rule of wine can be broken sometimes with good effect, but it's gotten me through many a fancy business dinner: white wine usually goes with white foods, and red wine goes with red ones. Think about it: chicken, fish, pasta with cream sauce? White wine. Steak, spaghetti, eggplant parm? Red wine.

The most seasoned wine connoisseurs love

be forced to tip 20 percent by the restaurant. Be sure to check the bill—I've double-tipped too many times to count, especially when there was Chardonnay involved.

When you order home delivery, you should tip 10 to 20 percent of the bill if the order is from a nice restaurant, or 10 to 15 percent if the delivery is informal, family-style food, depending on the size of the order and how difficult it is to transport to your home (say, if you live in a fifth-floor walkup, or if the driver needed to make more than one trip from his car).

If you're just having drinks, a good rule of thumb is a buck a drink at a casual bar, or 10 to 15 percent of the bar bill somewhere fancy.

to taste. So if you're not sure what you're looking for and the restaurant offers wines by the glass, ask to taste something. This is a great conversation starter with your date and can keep you from spending the price of a fabulous top on a bottle of wine you hate.

If you really don't know what you're doing and you're at the kind of restaurant where the server's knowledge of wine is likely comprehensive and totally intimidating, give that garçon a chance to shine. There's nothing a waiter enjoys more than being asked, "What do you suggest?" And turning the floor over to a pro shows your date you're confident, friendly, and always willing to learn.

You've just been designated your table's wine taster—but you don't know what the wine is supposed to taste like!

I once went to dinner on a blind date with a man who ordered—and sent back—four different bottles of wine before he settled on one he could tolerate. Many people don't realize that when a server or sommelier asks you to taste the wine you've ordered, it's to ensure that the wine hasn't spoiled, not that it's to your liking. "Corked" wine means that the phenols in the cork react with the fungi and chlorine in the cork (chlorine is used to sanitize the cork), producing a chemical called *trichloroanisole*, or TCA. Wines in contact with a "contaminated" cork will end up tasting bad. It's estimated that up to 5 percent of wines bottles are corked. To make sure yours isn't one of them, pick up the cork and smell it. As long as it doesn't smell musty or moldy, pick up your glass and sniff the wine. It shouldn't smell "off," either—and believe me, you'll know if it's off. Some people describe the smell of corked wine as reminiscent of wet dog or indoor swimming pool, which makes sense considering that TCA is a chlorine compound. Then take a swig and swirl it around in your mouth. Remember, you're just making sure it's not corked—not offering an opinion on taste. If you think it's off but are unsure, don't hesitate to ask your server or sommelier for his or her opinion. That's what he or she is there for.

Your credit card is declined.

First, ask yourself if you've done something that might make your server want to punish you. I once knew a waiter who told me that when people are exceptionally rude to him, he gets them back by embarrassing them in front of their date.

If you think the matter is truly financial, apologize and defuse any tension by saying, "I've been having some problems with this card lately—will you excuse me while I make a quick call?" Step outside and call the number on the back of your card to find out what's behind the company's refusal to accept the charge. If you've gone just over your limit, the customer service representative may be able to push the charge through, especially if you have a good payment record. Don't be afraid to tell them they are in a position to save you from dying of humiliation.

If your card just plain isn't going to work and you don't have another payment option, tell your companion that you're mortified but you're going to need to borrow your share. If he or she insists on paying the bill in full, say that dinner's on you next time—and set a date so the other person knows you're serious.

How to Eat an Oyster Like a Lady

Don't let these difficult foods get the best of you.

- **Oysters.** Think of an oyster like a shot. Put it up to your lips, tip your head back, and swallow. Try not to dribble, and make eye contact with your date to draw focus away from any mishaps toward the windows to your soul.
- **Lobster.** Deconstruct a lobster as you would an enemy. Rip off its claws first, crush them with a cracker and pull out the meat. Then bend the body back from the tail, and pull out the black vein. Use your dainty little fork to dip the meat in clarified butter. And don't be shy about wearing your bib.
- **Soup.** Always wait for soup to cool enough for you to eat delicate mouthfuls comfortably and quietly. Scoop your spoon away from you, toward the back of the bowl, to avoid splashing yourself.

At a big group dinner where everyone has snarfed down steak, lobster, and Château Margaux while you've crunched a salad and washed it down with lemon water, you're asked to pay an equal share of the check. You don't want to be cheap, but this isn't fair.

First of all, whom are you dining with? If it's a casual dinner with a group of close pals, just say something like, "You know I love you boozers and gourmands, but until I sell that screenplay I can't afford to keep you in white truffles. Here's my $20."

If it's a group you're not as familiar with, try saying, "I apologize, but I didn't realize we were going for an even split, and since I only brought $20, I ordered the green salad. Hope a 40 percent tip is enough!"

Certain situations, such as birthday parties, family-style meals, or fancy double dates, necessitate an even split. In these cases, just be sure to order something you really enjoy so you're getting your money's worth.

Your date—or his parents—want you to try a food you hate, and you don't want to seem difficult or picky.

It's amazing how love can deaden one's taste buds as well as one's sense of logic. I was once dating a guy who surprised me with a tray of beluga caviar garnished with chopped eggs and raw onion. These are quite possibly my three least favorite foods, but I liked him so much that when he brought a toast point up to my lips as though I were a three-year-old, I opened right up like a good little toddler. I expected to gag, but I found the combo bizarrely delicious. While the relationship didn't last, my love for caviar service did.

Still, nine times out of ten, you think you hate a food because you actually do hate it. And when your boyfriend's mom begs you to try her gefilte fish, you need to make a decision: risk gagging (which will humiliate her) or refuse to try it (which will make you seem like a total bitch).

If you think you can stomach just one bite, do so, fake a smile, and hide the rest in your napkin when no one is looking. If there's no way you can get the food in question down, just say, "Forgive

Five Things You May Not Want to Order on a First Date

A plain salad, sans dressing (he'll think you're boring and weight-obsessed—are you?)

Spaghetti (too much potential for greasy tomato sauce stains, not to mention garlic breath)

Ribs (while the image of a woman gnawing on bones can be sexy, it can also go very, very wrong)

Corn on the cob (you don't want to look like a beaver)

An ice cream cone (sends the wrong message)

How to Ensure You Get Poor Service

1. Snap your fingers or wave wildly at the waiter to get his or her attention.
2. Yell across the room at any member of the restaurant's staff.
3. Let your kids run wild.

4. Let your baby throw food and don't bother picking it up.
5. Change your order once it's been put in.
6. Ask for two salads to be split between three people.

me, but I'm just getting over something and my stomach has been very sensitive lately. I really need to stick to simple foods right now. But it looks delicious!" (Not.)

Sticky Note: There are a number of foods that you are supposed to eat with your hands. So order asparagus, artichokes, crisp bacon, french fries, olives, raw vegetables, chicken wings—and make sure you have a lovely manicure.

On the Road

You've borrowed a friend's car, just for a day, and you nearly total it—well, at least it feels that way. Actually, it's just a scratch. But a really big one.

First of all, be grateful that the huge truck you thought you were avoiding didn't total you. Once you've thanked the gods of driving, you need to deal with the wheel issue at hand. Cars, like fancy sunglasses and favorite sweaters, are no fun to borrow.

Step one: minimize the damage. Windex (or an eco-friendly version, such as Seventh Generation) and Bounty contribute more than you might imagine to erasing a car scratch. Often what looks like a scratch is really the other car's paint, and that comes right off.

If that doesn't work, take it to an expert: the guy at the car wash. He'll buff that scratch like nobody's business—especially the owner's—for a few extra bucks. If that fails, go to an auto body shop before you have to return the car. For $80 you might get that mirror repainted—and your sin washed clean.

Step two: manage your relationship. Your friend probably doesn't need to know you didn't learn to drive until after you lost your virginity, but if he finds that scratch without your telling him, you've lost a generous friend—and his car. Tell him you noticed a scratch and that it must have happened on your watch. You'd like to pay for it to be fixed by the dealer of his choice.

A slightly lamer—but still acceptable—path would be to ask if the scratch was already there when you borrowed the car, because you think it's new, but you're "not sure." Either way, mention it, take the high road, and then get your own car.

Your yoga friend always asks you for a ride to class. Granted, she's kind of on the way—kind of. You're sick of paying for the gas and having to include her in all your after-yoga plans.

Isn't the point of yoga to transfer some of that peace-loving, generous way of life from your mat into the real world? Get over it—not everything outside the yoga studio is fair and balanced. Make sharing the ride a friendly ritual that you look forward to. It's nice to have a pal to keep you going to class, so she's giving you a gift, too.

Next time, after class, ask if she wants to go out for an açai smoothie. You can enjoy your buzz, and my guess is she'll pick up the tab—I certainly would. Thus your $2.39 worth of gas is already covered, and then some. *Namaste.* (If she doesn't offer to buy you a smoothie when you've been serving as her own personal school bus all year, all bets are off, BTW.)

PART 10

Icky Situations

You can barely even admit this, but . . . you have a pimple. On your (shhhhhhhh) labia.

Yes, it's gross—but it happens to the best of us. And guess what—nine times out of ten, it's just a pimple. Treat it with warm compresses, and whatever you do, don't pop it. This could spread infection. If it doesn't vanish in a week or so, you should head to the doctor. He or she can drain it and put you out of your misery.

You have a boil.

Boils are abscesses—pockets of pus—under the skin. Wow. Gross. The main distinction between boils and pimples or cysts is that boils hurt. A lot. There are over-the-counter creams to help with pain relief, but the best way to alleviate the pain of a boil is to encourage it to drain by applying warm compresses. Never, ever try to coax a boil out of its hole, because popping can lead to serious infection. If you can't shake this nasty bump, your MD can lance it for you.

Dear Jane,
My vagina clenches up every time I try to have sex. I can't get a penis in there!

Sounds like you may have vaginismus, an involuntary tightening of the muscles of your pelvic floor. This can have both psychological and physiological causes. The good news is, the tightness is totally treatable. Check in with your gynecologist or therapist, who can help you figure out what's causing the problem. Then he or she will recommend a course of treatment, which can range from working through repressed sexual issues to slowly inserting phallic objects into your vagina under supervision.

You got a pedi yesterday, and now your foot is pink, tender, and the size of your face.

Lest you think an infection you got at a nail salon couldn't possibly be a big deal, consider that a Texas woman recently died from a staph infection acquired by a cut inflicted with a pumice stone. If you notice any signs of infection on your hands or feet—redness, swelling, tenderness, tightness, pus—head to the doctor immediately. To avoid contact with superbugs that can lurk in nail shops, make sure all instruments are sterilized in an autoclave and that footbaths are lined with hygienic plastic that's changed for each and every customer.

Your pee smells (and it's been at least six months since you last ate asparagus).

Lots of things can affect the smell of urine, from vitamins to onions and, yes, asparagus. If your pee has a strong smell, start by drinking more water—

it may simply be highly concentrated. If the smell doesn't resolve, head to the doctor just to be sure your kidneys are working properly.

You have halitosis.

In very rare cases, chronic bad breath can be an indication of a serious health issue. But more likely it can be resolved with a humble yet magical tool—the tongue scraper. Drag this U-shaped metal instrument from the back of your tongue to its front and you'll be amazed at the goop that comes off—and the stench that goes with it.

You are on the couch watching TV with your boyfriend and you really, really, really have to fart.

Stand up delicately, clenching your buttocks tightly, and head to the bathroom. If you can't make it, at least venture into another room. Should you have no choice but to let it rip, delicately say, "Excuse me." Trust me, he'll probably think it's adorable.

Afterword

Well, we've covered icky, awkward, embarrassing, messy, bothersome, banal, stupid, and downright sticky. There are a million more sticky situations I've stepped into or been asked to include (and if you have more, please send them to me at jane@moderngirlsguide.com or tweet me at twitter.com/jane_buckingham). But what I hope, my friendly reader, is that I've provided enough guidelines to get you through your next sticky situation with ease. Because a mess is only as sticky as you let it be, and no situation is too sticky to let today's Modern Girl *ever* come truly unglued.

With kind affection to all the
Modern Girls out there,
thanks for reading
JB

Acknowledgments

This book would not have been written without the time, efforts, diligence, and plain old talent of Rebecca DiLiberto. While I think there wasn't one before, I'm confident never again will there be a sticky situation you can't emerge from.

Victoria Doramus researched and dug and wrote and found and searched and looked all while doing *another* full-time job. We've been through so much together, and I thank you for all of your support through it all.

Cassie Jones, without you there would be no sticky, no situations, no Modern Girl and really no reason to keep writing. You are an editor of yesteryear, and every author should be lucky enough to have you by her side. Your continued support and friendship means so much more to me than I think you know.

Jennifer Rudolph Walsh, thank you for encouraging me to get out of that "studio apartment." You were right, as always. Your advice and support are always appreciated.

Andy McNicol, you listen, you think, you question, and you find the best solutions out there. You are the partner in these projects I could only hope for. Thank you so very much for everything you do for me. It wouldn't happen without you.

Stephanie Daniels, thank you for enduring my endless crazy requests. You are a lifesaver.

Jessica Deputato, thank you for working to keep this book on track. Day after day after day.

Michael Rinzler, thank you for your love, and for always getting me out of familial sticky situations. Now more than ever will show up.

Joni Evans, you are more than a legend, mom, and oracle. How did I get so lucky to have you in my life?

Linda and Mitch Hart, thank you for always being there when times were and are the stickiest.

Jo Boorman and Pippa Duke, your love and faith are incredible. Thank you.

Paul and Dad, thank you for showing me the spiritual side.

Julie, Mark, and Luke Rowen, thank you for much needed support every Sunday, and every day in between.

Kate White, thank you for inspiring me, for the gift of friendship, and for never letting me forget my mother.

Sylvia Cerna and Olivia Sinaguinan, thank you for making my life run so much more smoothly so that I can write without fear that everything else will fall apart.

Marcus, thank you for always loving me, no

matter how prickly I get when things get sticky. I love you.

Jack and Lilia, thank you for being patient and always putting a smile on my face. Please remember, no one's problems are more important to me than yours. I love you so much.

To my girlfriends, whom I cherish beyond words: Allison Druyanoff, Amanda Freeman, Amanda Stefan, Andrea Simon, Andrea Stanford, Barbara Coulon, Candie Weitz, Clare Ramsey, Dana Oliver, Dawn Ostroff, Debbie Meizenzahl, Elizabeth Harrison, Elisabeth Hasselbeck, Jackie Smith, Jenny Levin, Julia Sorkin, Katie Tarses, Lauren Samuel, Kristin Bennett, Liz Gray, Melissa Lavigne, Melissa Thomas, Nancy Paul, Nina Kotick, Norah Weinstein, Rebecca McQuigg, Soleil Moon Frye, Susan Smidt, Stacy Twilley, and Trisha Cardoso. Without all of you in my life, I could not be a mom, writer, wife, and friend, and stay sane. Thank you, truly, for forgiving my failures and supporting me.

To the girlfriends I adore and stupidly forgot to name above, it is *not* that I don't love you . . . I am just dreadful at writing these things and always leave *someone* out (hence, the book).

Dina Gerson, I marvel at your strength over the last year.

And finally, thank you to two women who have endured journeys that go far beyond "sticky." Andrea Hutton and Sarah Geary, your strength, courage, and grace are remarkable. You are inspirations whether you like it or not. I am grateful to have you as friends.

Index

ALSO BY JANE BUCKINGHAM

THE MODERN GIRL'S GUIDE TO LIFE

ISBN 978-0-06-073416-9 (paperback)

"A hot pink how-to manual for chic career girls, packed with powerful advice—from party planning and 'wediquette' to shopping, packing and 'Makeup Tips for Busy Chicks.'"

—*Us Weekly*

THE MODERN GIRL'S GUIDE TO MOTHERHOOD

ISBN 978-0-06-088534-2 (paperback)

The perfect book for the young mom, including the best-kept secrets, practical advice, and multiple solutions for problems from pregnancy and birth to age five.